AVIATION GROUND OPERATION

National Safety Council
1121 Spring Lake Drive
Itasca, IL 60143-3201

SIXTH EDITION

Executive Director of Publications:	Suzanne M. Powills
Book Editor:	Phyllis Crittenden
Technical Adviser:	Barbra Jean "B.J." Lo Mastro
Cover and Interior Design and Composition:	Jennifer Villarreal

Library of Congress Cataloging-in-Publication Data

National Safety Council.
  Aviation ground operation safety handbook. -- 6th ed.
     p.cm.
  Includes bibliographical references and index.
  ISBN 978-0-87912-266-9
     1.Airports--Safety measures--Handbooks, manuals, etc. 2.
Aeronautics--Safety measures--Handbooks, manuals, etc. I. Crittenden,
Phyllis. II. Title.

TL553.5.N32 2007
629.136028'9--dc22
.1C0917   2007014379

Contents

INTRODUCTION

The International Air Transport Section of the National Safety Council is excited to support the publication of the sixth edition of the Aviation Ground Operation Safety Handbook. The International Air Transport Section (ARTEX), a Section of the National Safety Council's Business and Industry Division with international membership, has been working diligently to produce best practice documents that will be a valued addition to this handbook. Within the past few years, ARTEX members have made a commitment to complete working documents within a brief prescribed time frame. These documents are designed to assist both the new aviation safety professional as well as those who are well seasoned in addressing the complexity of issues related to aviation ground safety.

Although many safety principles are common throughout all industries, there are unique issues related to aviation ground safety that apply around the world. By tapping into our international connections, world-class ground safety practices have emerged. Whether the issues are related to maintenance operations like fuel cell entry, loading passenger bags into a cargo hold, handling aircraft during severe weather situations or preventing employee falls from aircraft wings this edition has been updated to include the concepts that will support worker safety throughout the industry.

As the first female General Chairman of ARTEX since its inception in 1928 it had been my honor to serve in a leadership role among such dedicated and knowledgeable safety professions. It is the hope of the International Air Transport Section that the newest best-practice papers in this sixth edition will help to improve the overall ground/occupational safety of airlines and their employees around the world.

TERRI WEILAND, GENERAL CHAIR
TR WEILAND COMPANY, LLC

ECONOMIC IMPACT OF INCIDENTS

Safety in aviation ground operations is an integral part of maintaining safe, reliable, economical airline service. Safety practices in the shop, hangar, and on ramps, and in the terminal may very well determine the safety of passengers in the air. This handbook suggests practices and programs that will improve ground safety.

The employee, family, and society suffer as a result of an accident. There is also significant economic effect on a company's profits. Many line management personnel are under the false impression that all losses resulting from an accident are covered by insurance. In fact, the indirect and usually uninsurable costs can be many times higher than the direct costs. Some of the indirect costs are these: the cost of overtime or additional wages to replace an employee; the cost of time spent by supervisory personnel in investigation and preparing required accident reports; and the cost of selecting and training a replacement employee. If an aircraft is damaged, costs may increase because of loss of passenger revenue, ferry expenses, flight crew salaries, and so on. Insurance premiums are usually based on past incident experience and also increase as the losses increase.

SAFETY POLICY

Management's responsibility for preventing accidents is the same as its responsibility for administering any other business function. A high incident frequency is an indication of inadequate management policies and safety programs. Supervisors must be held responsible for accidents in the same manner as for on-time performance and cost control.

A statement of the safety policy that clearly defines the responsibilities of management and employees in maintaining a safe working environment should be published and enforced.

EFFECTIVE GROUND SAFETY ADMINISTRATION

The administration of a safety program may be delegated to one person, preferably a qualified health and safety professional; however, the ultimate responsibility for preventing incidents still rests squarely with management.

The person selected to administer an accident prevention program should report to a top executive officer in the company. In this position, the safety and health professional will be able to advise and to represent management.

Companies that do not have an assigned individual for administering the accident prevention program should seek assistance from their insurance carriers, the National Safety Council, and the appropriate governmental agencies.

HANDBOOK SCOPE

This handbook is designed to guide the user in the prevention of accidents that may result in injuries/illnesses to employees and the general public, and in damage or other loss to customer and company property or equipment.

This handbook sets forth the guidelines for safely accomplishing most ground operations associated with aircraft. These guidelines are applicable to aviation ground operations. Such as:

Air carriers

Fixed base operators

Corporate fleet operators

Airport operators

Aircraft service contractors.

In addition to the material in the text, the reference sections at the end of some chapters list many codes, standards, and other publications. It is strongly recommended that readers refer to these codes or standards in their jurisdiction to obtain more information.

SIXTH EDITION

This sixth edition is an update by the Section Members of the International Air Transport Section of the National Safety Council's Business and Industry Division. Its express purpose is to give practical guidelines that will help reduce ground accidents within the industry. The contents are not regulations but suggested good practices. Where any conflict exists between these suggested practices and the legislation in the country where they are to be applied, the legislation will, of course, take precedence.

THANKS TO...

We extend our thanks to the section member whom contributed to the publication of this safety handbook. A special thanks to Kevin W. Frommelt, Northwest Airlines, Chair of the Aviation Ground Operation Safety Handbook Project. The National Safety Council would also like to express special thanks to David Marquette, Marsh, John Kane, Air Canada, Jim Swartz, Delta Air Lines, Jim Stephan, Delta Air Lines for their time and dedication on the final review and analysis of this manual. A special thanks goes to B.J. LoMastro, Section Manager, Business and Industry Division.

SAFETY MANAGEMENT SYSTEMS

CHAPTER 1

Safety management systems
Continuous improvement cycle
- Policy
- Planning
- Implementing
- Measuring
- Reviewing and improving
- Summary

Safety programs
- Management leadership and employee participation
- Worksite analysis
- Hazard prevention and control
- Training and communication
- Summary

Support systems
- Organizational structure
- Documentation
- Audits and periodic reviews
- Organizational culture
- Summary

References

A safety management system is a systematic and comprehensive approach to accident prevention, and consists of a number of interdependent parts. When organizations implement a safety management system they are in a better position to:

- Prevent accidents
- Meet legal and moral obligations
- Attain business objectives

Successful safety management systems achieve success through a cycle of continuous improvement and share the following basic characteristics:

- A requirement for the establishment of policy, goals and objectives
- The development and implementation of safety programs to achieve these goals and objectives
- A mechanism to continuously measure performance and make improvements
- The conducting of audits and periodic reviews to ensure compliance and system effectiveness

Organizations also must recognize that their safety management system is supported and influenced in many ways. To ensure optimum results, it is necessary to give careful consideration to these influences when developing strategies for performance improvement. These influences have a significant impact on the overall effectiveness of your effort.

A safety management system has three distinct components:

1. Continuous improvement cycle
2. Safety programs
3. Support systems

Understanding the role and interdependencies of these components will help you manage safety in your organization.

The International Organization for Standards (ISO) and the International Labour Organization (ILO) have established standards and guidelines on management systems, and are a good reference for organizations in the process of developing or seeking to improve safety management systems.

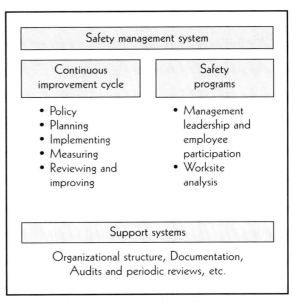

Figure 1-1. Components of a Safety Management System.

CONTINUOUS IMPROVEMENT CYCLE

The continuous improvement cycle is an important safety management system component. This cycle of continuous improvement provides the feedback loop and overall control in the system. For example: The continuous improvement cycle requires that organizations compare their actual performance against established objectives. When an organization fails to meet its performance targets, an analysis is conducted to understand why these objectives were not met. Once the reasons are understood, safety programs can be developed and implemented or improved, to address the concerns. The implementation of this cycle of continuous improvement is what drives excellence in safety performance.

Safety often is managed as part of the overall management system in an organization; it is not necessary to create a separate cycle of continuous improvement solely for the purpose of managing safety. Many organizations have integrated this activity successfully within an existing management system. The important point is to establish a continuous improvement cycle that consists of the following activities: policy, planning, implementing, measuring, reviewing and improving.

Policy
The establishment of a safety policy is the first order of business. The chief executive officer must establish clear expectations for safety performance. A safety policy must demonstrate the organization's commitment to safety. This policy statement will provide guidance and direction for all employees. Information about the organization's safety mission, vision, values, principles, goals, objectives, performance targets, responsibilities and strategy must be communicated clearly to all employees. How you choose to communicate this information to your organization is not the critical factor. Some organizations choose to include all this information in one safety policy statement, while others choose to establish a short concise policy statement that is supported with additional communications for the remainder of the information. The establishment and communication of a safety policy is vital to the success of your safety effort.

Planning
Once an organization has established a policy, and set its goals and objectives, it must plan the activities and strategies necessary to attain them. This safety plan acts like an organizational road map; it will define where you are going, how you will get there, who will be responsible, what resources will be involved, and so on. When developing a safety plan, an important consideration will be to ensure compliance with all applicable regulations. However, the scope of a safety plan should not be limited to regulatory compliance. Performance-based safety plans often encourage action beyond the compliance threshold – with excellent results.

During planning sessions, use a risk management approach: hazard identification, evaluation and control. This approach is the most commonly used in industry today and will facilitate decision-making. Risk management helps select the most appropriate and cost-effective solution to safety concerns. Successful safety management systems have integrated this approach into the overall system – usually as an organizational principle.

A final note on planning: It is important to consider how to solicit input from employees. Many organizations have had success using a team-based approach (e.g., safety and health committees) to create an environment of cooperation with consensus-based decision-making.

Implementing
Organization must provide the necessary resources – human and capital – to implement the activities in the safety plan. Failure to provide adequate resources will suggest a lack of commitment. It is important to understand this implication when developing a safety plan. Setting clear boundaries and involving employees during the planning phase will facilitate the process.

There is no one right way to develop an implementation strategy; however, there are two important considerations. First, the responsibility for implementation always will be a line function – with the safety staff providing assistance. Second, never underestimate the cultural influences within an organization when preparing to implement a safety plan; dialogue and input from employees is essential. Employees often can tell if strategies will work before the strategies are implemented.

Measuring

The saying "What gets measured gets managed" illustrates the importance of properly measuring safety performance. Particular consideration must be given to selecting the right things to measure when establishing this part of the overall system. Selecting what to measure in order to improve safety performance requires careful thought to make certain the desired actions are achieved. Performance measures are intended to provide feedback on the effectiveness of the organization's safety plan and focus attention on what the organization wants improved.

There are essentially two types of safety performance measures: rates and raw numbers. Both measurements are valuable but serve different purposes. Raw numbers occasionally are used when measuring significant or infrequent events, but often are problematic because they do not account for fluctuations in workforce numbers and other variables that affect actual performance. For example, if the number of lost workday incidents is chosen as an indicator, improvement can result due to personnel reductions, without actual safety performance improvement. Therefore, rates per employee, hours worked or employee hours generally are better used to gauge performance and identify positive or negative trends.

Safety performance measures are lagging indicators because accidents, injuries and illnesses are counted after the fact. Using lagging indicators is like looking in the rear view mirror when driving; a person cannot do much about what has happened already. Because of this, many individuals argue that organizations need to make more use of leading indicators – those that might allow intervention before an accident happens.

In theory, this sounds like a good idea, but it is not always easy to identify those things that will serve as reliable precursors to accidents. Nevertheless, with some effort, organizations can identify those actions that, if not properly attended to, may have a negative impact. Communicating safety performance in business terms, such as: lost workdays, costs and/or production delays are especially useful in sensitizing employees to the impact that safety can have on business performance and may be necessary to drive safety improvements.

Reviewing and improving

Safety performance reviews must be given the same level of prominence as other business metrics, such as on-time performance, quality performance and financial performance. These reviews must take place on a regular basis and at all levels of the organization. Organizations that review safety performance on a regular basis are in a better position to respond to changes in performance by taking action and adjusting their plans accordingly. Finally, if managers and supervisors are expected to take safety seriously, then the level of accountability for poor safety performance must resemble – and be handled in the same manner as – any other business metric.

SAFETY PROGRAMS

Safety programs are intended to help organizations manage safety risks. A safety program can take the form of a standard operating procedure, a safety process or a safety initiative. Safety programs are the activities an organization deems necessary to ensure safety risks are properly managed. All these activities share the objective of accident prevention.

Safety programs can be divided into four main categories: Management leadership and employee participation, worksite analysis, hazard prevention and control, and training.

Management leadership and employee participation

Management leadership and employee participation are critical in the development of an effective safety program. Management must regard safety as a fundamental value, and ensure it receives the same level of attention as other key operational activities. First-line supervision plays an essential role in this process.

Management leadership activities include, but are not limited to, the following:

- Clearly communicating safety policy and expectations
- Encouraging senior management involvement
- Taking responsibility and ensuring employees have authority to make safety decisions
- Holding all employees accountable for safety
- Providing the framework to encourage employee participation
- Involving safety personnel in decision-making

Employee participation provides the avenue through which each employee can express his or her own commitment to safety. This can be accomplished through the following:

- Safety and health committees
- Safety and health problem-solving groups
- Safety observations
- Hazard reporting/accident investigations
- Safety communications

Worksite analysis

Worksite analysis includes a variety of worksite observations and evaluations that identify unsafe conditions or unsafe practices or behavior. On an ongoing basis an organization must conduct inspections, observations, and data analysis in an effort to identify potential hazards. Suggested activities include the following:

- Processes for reporting hazards without fear of reprisal (e.g., an anonymous hotline)
- Periodic safety surveys/audits
- A mechanism to ensure that recommendations are addressed in a timely manner

- Routine hazard analysis activities (e.g., Job Safety Analysis)
- Accident investigation and follow-up
- Analysis of injury/illness data, including trend analysis
- Evaluation of new processes, materials, equipment, etc., to determine potential hazards before being placed in operation
- Communication of accidents and trends with analysis

Hazard prevention and control

Effective safety programs identify hazards and concentrate their efforts on prevention. Hazards can be addressed through:

Engineering controls, which are preferred, but not always feasible. The redesign of a workstation to incorporate ergonomic changes is an example of an engineering solution.

Administrative controls, which may involve limiting the time one is exposed to a hazard or developing a safe work procedure.

Personal protective equipment (e.g., eye, hearing and respiratory protection), which must be applied when hazards cannot be controlled through an engineering or administrative solution. All PPE must be subjected to a hazard analysis prior to use. An authorized representative ensuring the assessment is accurate and complete must sign the hazard analysis.

Examples for preventing and controlling hazards include the following:

- Documented and ongoing monitoring of maintenance of workplace equipment
- Ergonomics program that includes hazard assessment and action taken to eliminate or reduce risk
- Medical surveillance programs that include functional capacity evaluations to ensure employees are physically capable of performing assigned job function(s)
- Developing and communicating safe work procedures associated with each job function

- Accountability process, designed to improve safe behavior
- Checklists
- Statistical process control charts
- Emergency response procedures

Training

Training focuses on providing employees with the necessary skills to ensure they are capable of performing their jobs safely. This includes ensuring employees fully understand the expectations and responsibilities associated with their job function. Training for safety must be practical and easily understood. The training must confirm that the employee has achieved the required and expected level of understanding.

Training for safety must ensure:

- Employees understand their safety responsibilities with respect to their job functions.
- Employees are made aware of potential hazards, and the procedures put into place to eliminate or minimize risk.
- Employees know when personal protective equipment is required and how to use it.
- Employees know where to go for safety information.
- Employees are aware of their responsibility to report accidents, near misses and hazards in a timely manner.
- Employees are aware of their responsibilities relative to accident investigation and follow-up.
- Employees know how to report safety concerns.
- Employees know how to perform their jobs safely.

SUPPORT SYSTEMS

Organizational structure

All organizations must ensure that a philosophical alignment exists from the highest level to the frontline employee. The direction and methodology of an organization's safety functions must be well-known and enthusiastically supported by all stakeholders.

All employees must understand their respective roles and responsibilities as they pertain to occupational safety. It is recognized that safety is everybody's responsibility; however, managers have the responsibility of ensuring their employees are protected at work. This responsibility is a line function and cannot be delegated. Staff functions such as safety managers are resources that line functions consult when seeking solutions to their safety concerns. This distinction is important – successful organizations understand this relationship.

Organizations need access to proactive individuals capable of driving safety improvement and assisting line functions with the resolution of safety concerns. Competencies include, but are not limited to: industrial hygienists, ergonomists, system safety specialists, safety managers and engineers. These skills can be employed within the company or contracted from other organizations.

Safety departments can exist as centralized, stand-alone entities reporting to the chief executive officer; or as decentralized departments within a division such as production, human resources, legal and risk management. Each approach has advantages and disadvantages. The important consideration is to ensure the safety department and its resources maintain some autonomy so that it remains true to its intended purpose: to maximize the effectiveness of the organization's safety management system without feeling compelled to compromise safety because of operational pressures.

Documentation

To implement and sustain continuous improvement, current practices must be documented. In addition to being a regulatory requirement in many countries, documentation is the necessary first step in establishing safety conformity in an organization. This is particularly important in large multi-site organizations.

A documented safety management system accomplishes the following:

- Communicates organizational expectations as they pertain to safety management

- Establishes the benchmark for audit compliance
- Serves as a training reference
- Is a repository of safe work procedures
- Fulfills legal obligations

Safety management systems can be documented using a traditional paper form or electronic media. Regardless of the type of documentation you choose, it is imperative you keep the documentation up to date, ensure that all employees have access to the necessary information, and develop a documentation format that is easy to navigate and understand.

Audits and periodic reviews

Audits and periodic reviews are a necessary part of any safety management system. Two activities must take place:

1. Audits to evaluate current organizational practices against regulatory requirements and company policy
2. Periodic reviews to assess the effectiveness of your safety management system

Compliance audits serve to identify nonconformance. In addition to being a due-diligence activity, compliance audits remind employees that they have a legal and moral responsibility to prevent personal injury and damage. A compliance audit identifies noncompliance to regulations and company policy by work location and is an indicator of an organization's commitment to safety. Corrective action plans must be established, and a system to follow up on noncompliance items ensures they are addressed.

Performance audits are closely related to the continuous improvement cycle. The primary focus is individual and organizational performance, as compared to the established goals and objectives. Audits can be used as a tool to build awareness and expert-ise in the organization. Supplementing audit teams with operational managers and employees encourages increased safety compliance and understanding.

Periodic reviews involve assessing the organization's safety management system. This activity usually is conducted every few years and involves input from various employee groups in the organization. The input provided allows the organization to identify strategies for continuous improvement of the safety management system.

Organizational culture

Organizational culture is a combination of the values, beliefs, attitudes and norms that exist in the organization. The failure to develop an effective safety culture is identifiable by the presence of at-risk behaviors. At-risk behaviors are a factor in the vast majority of personal injuries and damages. These behaviors can be found at all levels of the organization.

Organizations must create a culture that reduces at-risk behaviors and encourages safe work practices. One of the drivers in creating this culture is to ensure there are clear expectations established for safety. For this to happen, the organization's leadership must embrace safety as a core value and set standards of acceptable and unacceptable behavior. The organization's safety culture plays an important role in influencing the safety management system. This culture is fostered from the very top of an organization. All employees, including managers, must be held accountable in adhering to these expectations. The use of key performance indicators and performance management tools can help this endeavor succeed.

REFERENCES

Asfahl, C. Ray. *Industrial Safety and Health Management, 4th ed.* 1999.

Australian Standard *AS 4801*

Australian/New Zealand Standard *AS/NZS 4804*

Brown, Stephen; Don Jones; CSP, P.E.; and Hilary Paska. *VPP: Models of Safety and Health Excellence.*

Geller, Scott *E. Working Safe: How to Help People Actively Care for Health and Safety.* 1998.

International Air Transport Association (IATA). *Airport Handling Manual, 24th ed.*

International Labour Organization. *ILOOSH 2001, Guidelines on Occupational Safety and Health Management Systems.* 2001.

International Organization for Standards. *ISO 9000.*

Lack, Richard W. *Essentials of Safety and Health Management.* 1996.

Maurino, D. E., Reason, J., Johnston, N., Lee, R.B. *Beyond Aviation Human Factors.* 1995.

Noble, Michael T. *Organizational Mastery with Integrated Management Systems: Controlling the Dragon.* 2000.

Pybus, Roger. *Safety Management: Strategy and Practice.* 1996.

Senge, Peter. *The Fifth Discipline Fieldbook.* 1994.

Transport Canada. *Safety Management Systems, A Guide to Implementation.* 2002.

U.S. Occupational Safety and Health Administration (OSHA). *Safety and Health Program Management Guidelines.*

Waring, Alan. *Safety Management Systems.* 1996.

WORKING GROUP MEMBERS

Chairman:

John Kane – Air Canada

Co-Chairman:

Tim Racicot – Continental Airlines

Members:

Tom Dyce – Air Canada

Ron Failing – Frontier Airlines

Maxwell Fogleman, Ph.D., MPH – Embry-Riddle University

Choon Woo Lee – Korean Air

Michael Lueck – ABXAir Inc

David Marquette – Marsh Risk Consulting

Barry Murphy – Cullinagh Newcastle West Co.

Dr. Bertha E. Rodriquez – Mexicana Airlines

Robert Schwartz – Air Canada

Contributors:

Douglas F. Briggs – Boeing Commercial Airplane Group

Willy Pfister – SP Techinics, TEB

John Phillips – United States Air Force

Thomas Schermesser – Air France

Alejandro Luna Sotura – Aero Mexico

James E. Swartz – Delta Airlines

David Thomson – Southern California Safety Institute

Richard DeBerger – Airbus

Bill Simpson – Piedmont Hawthorne

RISK MANAGEMENT

CHAPTER 2

Risk management takes aviation safety to the next level. It is a six-step, logic-based approach to making calculated decisions on human, material and environmental factors before, during and after operations (Figure 2-1). Risk management enables senior leaders, functional managers, supervisors and others to maximize opportunities for success while minimizing risks. Failure to successfully implement a risk management process will have a financial, legal and social impact.

Why risk management?
Financial impact
- Achieves financial objectives
- Controls indirect costs
- Preserves resources
- Ensures business continuity
- Enhances business/community reputation
- Sustains positive business/customer perceptions.

Social impact
- Maintains obligations to employees
- Provides moral leadership
- Improves communication at all levels
- Increases employee morale

Legal impact
- Ensures regulatory compliance
- Avoids civil and criminal penalties

Work processes
- Boosts productivity

- Supports timely/correct delivery of services/ product
- Improves quality
- Increases reliability
- Provides consistency

Customer service
- Improves customer retention.
- Expands market share

Risk management principles
The following three principles govern all actions associated with risk management. These continuously employed principles apply before, during, and after all tasks and operations.

1. Accept no unnecessary risk. Unnecessary risk comes without a commensurate return in terms of real benefits or available opportunities. All aviation operations and related activities involve risk. The most logical choices for accomplishing a task are those that meet all requirements with the minimum acceptable risk.

2. Make risk decisions at the appropriate level. Making risk decisions at the appropriate level establishes clear accountability. Those accountable for the success or failure of the operation must be included in the risk decision process. The appropriate level for risk decisions is the one that can allocate the resources to reduce the risk or eliminate the hazard and implement controls. Typically, the person responsible for executing the operation or task is:

- Authorized to accept levels of risk typical of the planned operation.
- Required to elevate decisions to the next level in the leadership chain after it is determined that controls available to him/her will not reduce residual risk to an acceptable level.

3. Integrate risk management into operations and planning at all levels. Risks are more easily assessed and managed in the planning stages of complex operations. Integrating risk management into planning as early as possible provides

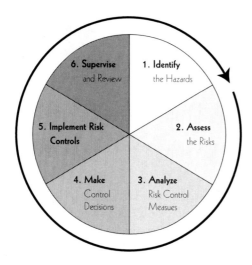

Figure 2-1. The pie chart shows the six-step logic-based, commonsense approach to making calculated decisions on human, material, and environmental factors before, during and after operations.

the decision-maker the greatest opportunity to apply risk management principles. Additionally, feedback must be provided to benefit future missions/activities.

THE RISK MANAGEMENT PROCESS

Risk management is a continuous process designed to detect, assess, and control risk while enhancing performance and maximizing effectiveness. The specific actions associated with each step of the risk management process are depicted in Figure 2-2.

Step 1. Identify the hazards

The purpose of Step 1 is to identify as many hazards as possible. A hazard can be defined as any real or potential condition that can cause injury, illness, death to personnel, or damage to or loss of equipment or property.

A list of common types of aviation-related hazards is provided in Appendix 1.

Additionally, the following tools can be used to help identify additional hazards.

Operations analysis
Purpose: To understand the flow of events.

Method: List primary events/actions in sequence. Use time checks if possible. The list of generic aviation-related activities in Appendix 2 can help identify tasks associated with several areas.

Hazard analysis
Purpose: To get a quick hazard survey of all phases of an operation. In low-hazard situations the preliminary hazard analysis may be the final hazard ID tool.

Method: Tie it to the operations analysis. Quickly assess hazards using scenario thinking, brainstorming, experts, accident data and regulations. This method considers all phases of the operation and provides early identification of highest risk areas. It also helps prioritize areas for further analysis.

"What if" analysis
Purpose: To capture the input of personnel in a brainstorming-like environment.

Method: Choose an area (not the entire operation), get a group of personnel involved together and generate as many "what if" scenarios and outcomes as possible.

Scenario process tool
Purpose: To use imagination and visualizations to capture unusual hazards.

Method: Often called the "mental movie" technique, a person visualizes the flow of events and injects "Murphy's Law" at every possible event. Use the operations analysis as a guide when imagining what should take place.

Change analysis
Purpose: To detect the hazard implications of both planned and unplanned change.

Method: Compare the current situation to a previous situation to identify any new hazards.

Step 2. Assess the risks
Risk is the probability and severity of loss from exposure to the hazard. The assessment step is the

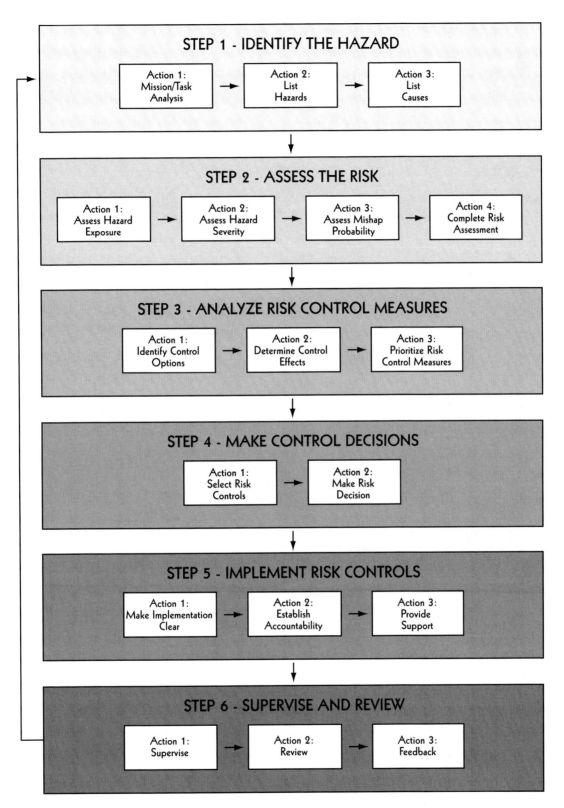

Figure 2-2. The risk management process.

application of quantitative or qualitative measures to determine the level of risk associated with a specific hazard. This process defines the probability and severity of an undesirable event that could result from the hazard. When conducting this assessment, don't overlook the information available that might help to quantify risks beyond your personal experience or intuition. A list of these types of information sources can be found in Appendix 3.

The Risk Assessment Matrix is a very useful tool in categorizing the effects of probability and severity as they relate to risk levels (Figure 2-3). To work through the matrix, first determine the severity of the hazard in terms of its potential impact on the people, equipment or task. Severity assessment should be based upon the worst possible outcome that can reasonably be expected. Hazard severity categories are defined to provide a qualitative measure of the worst credible mishap resulting from personnel error; environmental conditions; design inadequacies; procedural deficiencies; or system, subsystem, or component failure or malfunction. The potential severity of a mishap can be categorized as the expected injury/disability to personnel or the expected damage to equipment.

The next step is to determine the probability that the hazard will cause a mishap. Probability may be determined through estimates or actual numbers, if they are available. A qualitative hazard probability may be derived from research, analysis, and evaluation of historical safety data from similar tasks and systems. The probability of occurrence should refer to the probability of an incident/consequence as opposed to the probability of an individual hazard/basic event occurring.

Finally, by combining the estimates for severity and probability, the intersection of the appropriate row and column on the matrix indicates the risk level for the hazard assessed.

Severity

Catastrophic (I): Death or permanent total disability, system loss, major property damage

Critical (II): Permanent, partial, or temporary total disability in excess of 3 months, major system damage, significant property damage

Marginal (III): Minor injury, lost workday incident, minor system damage, minor property damage

Negligible (IV): First aid or minor medical treatment, minor system impairment

Probability

Frequent (A): Occurs often, continuously experienced

Likely (B): Occurs several times, occurs often

Occasional (C): Occurs sporadically, occurs sometimes

Seldom (D): Remote chance of occurrence; unlikely, but could occur at some time

Unlikely (E): Can assume it will not occur

Risk levels

Extremely high: Loss of ability to accomplish task

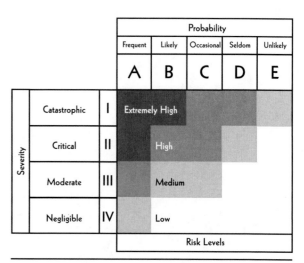

Risk Assessment Matrix

Figure 2-3. The Risk Assessment Matrix.

High: Significantly degrades capabilities in terms of required standards

Medium: Degrades capabilities in terms of required standards

Low: Little or no impact on accomplishing task

Residual risk: Risk remaining after risk reduction efforts.

Step 3. Analyze risk control measures

Investigate strategies and tools that reduce, mitigate, or eliminate risk. Effective control measures reduce or eliminate one of the three components (probability, severity, or exposure) of risk.

Macro options

Reject: The organization can – and should – refuse to take a risk if the overall costs of the risk are too high.

Avoid: Avoiding risk altogether requires canceling or delaying the job, mission or operation.

Delay: It may be possible to delay a risk. If there is no time deadline or other operational benefit to speedy accomplishment of a risky task, then it often is desirable to delay the acceptance of risk. During the delay, the situation may change and the requirement to accept the risk may go away.

Transfer: Risk transference does not change probability or severity of the hazard, but it may decrease the probability or severity of the risk actually experienced by the individual or organization accomplishing the operation/activity. As a minimum, the risk to the original individual or organization is greatly decreased or eliminated because the possible losses or costs are shifted to another entity.

Spread: Risk is commonly spread out by either increasing the exposure distance or by lengthening the time between exposure events.

Compensate: Redundant capability can be created

in certain special circumstances. An example is to plan a back-up; then if a critical piece of equipment or other mission asset is damaged or destroyed capabilities are available to bring online to continue the mission.

Reduce: The overall goal of risk management is to plan operations or design systems that do not contain hazards. A proven order of precedence for dealing with hazards and reducing the resulting risks is:

1. Plan or design for minimum risk. Design the system to eliminate hazards. Without a hazard there is no probability, severity or exposure.

2. Incorporate safety devices. Reduce risk through the use of design features or devices. These devices usually do not affect probability but reduce severity: For example, an automobile seat belt doesn't prevent a collision but reduces the severity of injuries.

3. Provide warning devices. Warning devices may be used to detect an undesirable condition and alert personnel.

4. Develop procedures and training. Where it is impractical to eliminate hazards through design selection or adequately reduce the associated risk with safety and warning devices, procedures and training should be used. The following options assist in identifying potential controls:

- Engineer
- Train/educate
- Guard
- Warn
- Improve task design
- Motivate
- Limit exposure
- Reduce effects
- Selection of personnel
- Rehabilitate

Step 4. Make control decisions

After controls have been selected to eliminate hazards or reduce their risk, determine the level of residual risk for the selected operation or course of action.

Accept the plan as is. Benefits outweigh risks (costs), and total risk is low enough to justify the proposed action if something goes wrong. The decision-maker must allocate resources to control risk. Available resources are time, money, personnel and/or equipment.

Reject the plan out of hand. Risk is too high to justify the operation in any form. The plan was probably faulty in some manner, or the objective was not that important.

Modify the plan to develop measures to control risk. The plan is valid, but the current concept does not adequately minimize risk. Further work to control the risk is necessary before proceeding.

Elevate the decision to the next level. The risk is too great for the decision-maker to accept, but all measures of controlling risk have been considered. If the operation is to continue, the next level in the decision-making hierarchy must decide if the mission is worth it, and accept the risk.

Make risk decisions at the appropriate level. Factors below become the basis of a decision-making system to guide leaders:

- Who will answer in the event of a mishap?
- Who is the senior person at the scene?
- Who possesses greatest insight into the full benefits and costs of a risk?
- Who has the resources to mitigate the risk?
- What level makes the most operational sense?
- What level makes these types of decisions in other activities?

Step 5. Implement risk controls

Once the risk control decision is made, assets must be made available to implement specific controls. Part of implementing control measures is informing the personnel in the system of the risk management process results and subsequent decisions. Careful documentation of each step in the risk management process facilitates risk communication and the rational processes behind risk management decisions.

- Make implementation clear.
- Establish accountability.
- Provide support.

Step 6. Supervise and review

Risk management is a process that continues throughout the life cycle of the system, operation, or activity. Leaders at every level must fulfill their respective roles in ensuring controls are sustained over time. Once controls are in place, the process must be periodically re-evaluated.

Supervise: Monitor the operation to ensure:

- Controls are effective and remain in place.
- Changes that require further risk management are identified.
- Action is taken when necessary to correct ineffective risk controls and reinitiate risk management steps in response to new hazards.
- Any time personnel, equipment or operations change, or new operations are anticipated in an environment not covered in the initial risk management analysis, risks and control measures should be revaluated.
- Successful performance of the operation or task is achieved by shifting the cost versus benefit balance more in favor of benefit through controlling risks.

Review: After assets are expended to control risks, a cost benefit review must be accomplished to see if risk and cost are in balance.

- Is the actual cost in line with expectations?
- How did control measures affect performance?
- Was a feedback system established to ensure the corrective or preventive action taken was effective?
- Was documentation available to allow a review of the risk decision process?
- What measurements were in place to ensure accurate evaluations of how effectively controls eliminated hazards or reduced risks?

Feedback: Feedback informs all involved as to how

the implementation process is working, and whether the controls were effective. Feedback can be in the form of briefings, lessons learned, cross-tell reports, benchmarking, database reports, etc.

APPENDIX 1. REPRESENTATIVE LIST OF HAZARDS

Physical hazards

Mechanical

- Unprotected moving machine parts
- Parts with hazardous surfaces
- Uncontrolled moving parts
- Movable transport or work equipment
- Falling/overturning objects
- Vibration
- Housekeeping
- Walking/working surfaces

Electrical

- Dangerous flow through the body
- Electric arcs
- Electrostatic charges
- Egress

Thermal

- Contact with hot surface
- Contact with cold surface
- Exposure to extremely high/low ambient temperatures

Chemical/biological hazards
- Gas
- Vapors
- Fumes, smoke, dust particles
- Liquids
- Solid materials
- Microorganisms/viruses
- Genetically altered organisms
- Allergic and toxic substances
- Bodily fluids

Environmental hazards
- Noise
- Radiation
- Heat/cold
- Lighting
- Meteorological (lightning, precipitation, tornado, hurricane)
- Air quality
- Cabin pressurization

Ergonomic hazard
- Workplace design
- People/anthropometrics
- Equipment (moving/fixed)
- Procedures
- Heavy work

Psycho-social hazards
- Time constraints
- Aggression
- Drugs and alcohol
- Reactions to abnormal situations
- Inexperience levels
- Circadian rhythm

APPENDIX 2. REPRESENTATIVE LIST OF AVIATION GROUND FUNCTIONS

1. Underwing

   a. Ramp operations/baggage loading
   b. Aircraft fueling
   c. Cabin service
   d. Lavatory and potable water service
   e. Baggage and cargo handling
   f. Catering service
   g. Line maintenance
   h. Deicing operations

2. Above wing

   a. Ticket counter
   b. Jet bridge operation
   c. Customer contact
   d. Cart, skycap and wheelchair personnel

3. Reservation and clerical personnel

   a. Load desk
   b. City ticket office

4. Maintenance operations

   a. Hangar and shop operations

5. Bus and other company drivers

6. Cargo facilities and mail

7. Flight kitchens

   a. Food preparation

8. Flight operations

   a. Operations control

9. Airport Security

   a. Transportation Security Administration (TSA)

APPENDIX 3. STATISTICS AND HISTORICAL REVIEW

1. Injuries (frequency/severity)
2. Lost workdays (rates/trends)
3. Aircraft damages/inadvertent slide deployments
4. Equipment damages
5. Near-miss reports
6. In-flight incidents (engines, hydraulics, electrical, etc.)

In any discussion of accident investigations, it is necessary to define the word "accident." An accident is any unexpected event that interrupts an operation. Accidents do not necessarily result in injury or damage; they are always preceeded by inadequate procedures, unsafe conditions or both. The purpose of an accident investigation is to determine the causes of an accident, develop corrective measures, and make recommendations for elimination of the causes.

HAZARDOUS MATERIALS AND
HARMFUL PHYSICAL AGENTS

CHAPTER **3**

Chemical hazards
Informing employees about hazardous materials
Evaluating hazardous materials
Control of hazardous materials
Personal protective equipment
Handling chemicals
Spills
Emergency information
Solvent storage
Solvent disposal
Compressed gases
**Other hazardous materials and harmful
physical agents**
Noise
Nonionizing radiation
Ionizing radiation
Explosive devices
References

Common hazardous materials used at airport facilities and their possible harmful effects will be discussed in this chapter. Safety and health procedures that can be employed to avoid personal injury will be stressed. (See Chapter 19, "Cargo Operations," regarding shipping hazardous materials.)

Hazards connected with materials and processes such as chemicals, solvents, compressed gases, and physical agents can be listed under four categories. These are listed in Table 3-A.

CHEMICAL HAZARDS

Potentially hazardous materials used in aviation ground operations include: paint strippers, cleaning solutions, brighteners, fluxes, insecticides, coating materials, resins, asbestos, solvents, and compressed gases. Processes such as plating, degreasing, stripping, cleaning, dipping, welding, plasma spray, soldering, aircraft skin repair, and auto body repair may generate toxic gases, vapors, mists, or dusts. Potentially hazardous chemicals that may be used in airline operations are listed in Table 3-B.

Informing employees about hazardous materials
Line management must ensure operators are aware of the presence and the nature of any hazardous material in products they use. It is important that management identify and follow the manufacturer's recommendations for the storage, handling, and use of all hazardous substances.

The sample format shown in Figure 3-1, the Material Safety Data Sheet (MSDS), generally is used internationally by chemical manufacturers or importers to provide this information. Normally, an MSDS will be provided with an initial shipment and with the first shipment after the MSDS update. If the MSDS is not provided with the shipment that has been labelled as a hazardous chemical, the distributor or employer must obtain one from the chemical manufacturer or importer as soon as possible.

In 1983, the U.S. Occupational Safety and Health Administration (OSHA) first required the Hazard Communication standard for the airline industry. The standard requires a written program establishing responsibilities for chemical labeling, proper employee training on health and safety hazards of chemicals, along with information on personal protective equipment, precautionary methods, and other controls. The U.S. Environmental Protection Agency (EPA) establishes standards for the control and disposition of hazardous materials and substances.

Some of the basic steps to follow in developing an internal program for the control of hazardous materials are listed below:

- Perform a chemical and hazardous substance inventory of every department.
- Request a Material Safety Data Sheet (MSDS) for every hazardous substance or material (Figure 3-1).
- Assess employees' needs for training and personal protective equipment.
- Analyze measures for the control of hazardous materials, along with the ventilation requirement for each area in which they are used.
- Order industrial hygiene surveys to be conducted when required.
- Provide appropriate training for employees.
- Give all user departments current MSDSs, updated every two years or as required by a work location regulation.

TABLE 3-A. FOUR CATEGORIES OF HAZARDS

Health hazards

Toxic effects from inhalation, ingestion, or absorption through skin
Oxygen deficiency
Dermatitis from repeated contact
Chemical burns from skin or eye contact with strong chemicals
Radiation effects, both ionizing and nonionizing
Biological effects from airborne or waterborne pathogens

Flammability hazards

Materials that burn under specific conditions
Flammable and combustible liquids, gases, or vapors
Flammable solids
Cryogenic materials

Instability hazards

Materials that may burn, explode, or produce hazardous gases upon contact with other
chemicals, or when destabilized with heat or shock
Materials that may decompose and support combustion when heated

Special hazards

Materials having unusual reactivity with water
Materials with oxidizing properties
Materials of radioactive character or having radioactive hazards

- Provide for scheduled retraining of all users.
- Supplement training with periodic reviews and departmental safety meetings of hazards and reactions of various chemicals and substances.
- Maintain records indefinitely for those undergoing training.
- Provide any contractors whose employees will be exposed to hazardous materials or substances with appropriate MSDSs, so they can do their training.
- Provide proper container labeling.
- Provide training – when required – for non-routine tasks.

Where local, regional or other international governmental laws provide for more stringent applications or controls, follow those standards or requirements for those locations.

Evaluating hazardous materials

In addition to the MSDS, several references are available regarding the toxicity and hazards of materials. (See, in References, National Safety Council's *Fundamentals of Industrial Hygiene,* Hamilton and Hardy, Sax, American Industrial Hygiene Association, Chemical Manufacturers Association, *Patty's Industrial Hygiene.*)

TABLE 3-B. COMMON HAZARDOUS SUBSTANCES

Chemicals

Acetone	1,2-Dichlorobenzene	Nitric acid
Alcohols	Fluorides	Perchloroethylene
Amines	Hydrochloric acid	Petroleum distillates
Aromatic hydrocarbons	Isocyanates	Phosphoric acid
Benzene	Isopropanol	Styrene
Butyl cellosolve	Methyl ethyl ketone	Sulfuric acid
Chlorinates	Methyl alcohol	Toluene
Chromic acid	Methyl butyl ketone	1,1,1-Trichloroethane
Phenol and cresylic acid	Methylene chloride	Trichloroethylene
Cyanides	Methyl isobutyl ketone	Xylene

Dusts and mists

Aluminum	Cobalt	Silica
Asbestos	Fibrous glass	Titanium
Cadmium	Hexavalent chromium	Wood
Caustics	Lead	Zinc
Chromium	Nickel	Zirconium

Final determination of whether a material or process presents a hazard to employees should be left to qualified chemists, industrial hygienists, safety engineers, or fire prevention specialists. Such professionals may be available within an airline's organization. Other sources for this information include insurance carriers and private consultants. Governmental departments of health or labor may provide free occupational health consultation.

A number of direct-reading and air-sampling instruments are available for measuring concentrations of mists, vapors, or gases in ambient air or in confined spaces. The instruments can be used to determine whether an environment is safe or requires further evaluation and corrective action. With proper training by qualified specialists, personnel on site would be able to use them.

Such instruments include color-indicator tubes for a large number of gases and solvents, combustible vapor detectors, oxygen meters, halogenated hydrocarbon meters, and dust collection and smoke tubes to evaluate exhaust systems.

Air-Sampling Instruments published by the American Conference of Governmental Industrial Hygienists provides an excellent discussion of instrumentation (see References). In addition, several professional periodicals provide information on such equipment, including *American Industrial Hygiene Association Journal, Safety+Health, Professional Safety,* and others.

Control of hazardous materials

The following is the preferred order of approach to controlling hazardous materials or processes (this order is an OSHA recommendation and may not be accepted worldwide):

1. Elimination: Change processes to remove the hazard.
2. Substitution: Provide a less hazardous material or process for the one used.
3. Engineering controls: These include ventilation, noise control, enclosure of process.
4. Administrative controls: Alter the length of time and frequency of exposure.

5. Personal protective equipment: This includes respirators and earmuffs, safety glasses, and gloves. This is the least desirable method and should only be used for temporary operations or where other methods are not feasible.

Asbestos and chemicals such as benzene and carbon tetrachloride should be prohibited from use. When their use cannot be eliminated, strict controls must be established. In most cases, satisfactory substitutes are available.

Personal protective equipment

When hazardous chemicals are handled, stored, or used, wearing personal protective equipment is necessary. Generally speaking, chemical safety goggles or face shields, acid proof aprons, rubber or nonreactive synthetic gloves, appropriate cartridge masks, or self-contained breathing units are the most common types used. Employees must be instructed in how and when to use this equipment as part of the Hazard Communication Training Program. This equipment should be maintained in a clean and sanitary condition, and inspected regularly for defects.

Personal protective equipment must conform to published standards. Its use should be under the control of someone knowledgeable in the subject. For more information on protective equipment, see, in References, American National Standards Institute (ANSI) *Practices for Respiratory Protection* and *NIOSH-Certified Personal Protective Equipment.* See also Chapter 13, "Personal Protective Equipment," in this book.

Handling chemicals

Chemicals come in bags, drums, cans and bottles. The most dangerous type of container is a glass bottle. Glass bottles of acid, for example, should be carried in a supporting container; acid carboys should be left in their wooden crates. Acid should be transferred from one container to another by a pump or carboy tilters. The mouth-siphoning technique never should be used.

Material Safety Data Sheet

May be used to comply with
OSHA's Hazard Communication Standard,
29 CFR 1910.1200. Standard must be
consulted for specific requirements.

U.S. Department of Labor

Occupational Safety and Health Administration
(Non–Mandatory Form)
Form Approved
OMB No. 1218-0072

IDENTITY *(As Used on Label and List)*	Note: *Blank spaces are not permitted. If any item is not applicable, or no information is available, the space must be marked to indicate that.*

Section I

Manufacturer's Name	Emergency Telephone Number
Address *(Number, Street, City, State, and ZIP Code)*	Telephone Number for Information
	Date Prepared
	Signature of Preparer *(optional)*

Section II — Hazardous Ingredients/Identity Information

Hazardous Components (Specific Chemical Identity; Common Name(s))	OSHA PEL	ACGIH TLV	Other Limits Recommended	% (optional)

Section III — Physical/Chemical Characteristics

Boiling Point		Specific Gravity (H₂O = 1)	
Vapor Pressure (mm Hg.)		Melting Point	
Vapor Density (AIR = 1)		Evaporation Rate (Butyl Acetate = 1)	

Solubility in Water

Appearance and Odor

Section IV — Fire and Explosion Hazard Data

Flash Point (Method Used)	Flammable Limits	LEL	UEL

Extinguishing Media

Special Fire Fighting Procedures

Unusual Fire and Explosion Hazards

Figure 3-1. The Material Safety Data Sheet (MSDS) format normally used by U.S. manufacturers and importers to provide information about a product. The form on these pages can be reproduced and used to request technical information from the manufacturer.

Section V — Reactivity Data

Stability	Unstable		Conditions to Avoid
	Stable		

Incompatibility (*Materials to Avoid*)

Hazardous Decomposition or Byproducts

Hazardous Polymerization	May Occur		Conditions to Avoid
	Will Not Occur		

Section VI — Health Hazard Data

Route(s) of Entry:	Inhalation?	Skin?	Ingestion?

Health Hazards (*Acute and Chronic*)

Carcinogenicity:	NTP?	IARC Monographs?	OSHA Regulated?

Signs and Symptoms of Exposure

Medical Conditions
Generally Aggravated by Exposure

Emergency and First Aid Procedures

Section VII — Precautions for Safe Handling and Use

Steps to Be Taken in Case Material Is Released or Spilled

Waste Disposal Method

Precautions to Be Taken in Handling and Storing

Other Precautions

Section VIII — Control Measures

Respiratory Protection (*Specify Type*)

Ventilation	Local Exhaust		Special
	Mechanical (*General*)		Other

Protective Gloves	Eye Protection

Other Protective Clothing or Equipment

Work/Hygienic Practices

☆ U.S.G.P.O 1986-491-529/45775

Dangerous chemicals should be transferred from receiving to storage on flatbed hand trucks, special carboy trucks, or pallets that have sideboards. Good judgment in handling is necessary to prevent mechanical damage to packages containing hazardous chemicals. It is advisable to have chemicals placed on skids for storage to prevent contact with moisture or chemicals on the floor and, if necessary, to facilitate cleanup of spills.

Acid carboys, both full and empty, should be vented periodically to relieve internal pressures.

Hazardous chemicals should be mixed and used according to information provided by the manufacturer, or as specified by written job procedures. Employees should be audited periodically to ensure they are following safe procedures. Engineering control equipment also should be reviewed periodically to ascertain that it is functioning properly.

Incidents often occur when reactive chemicals or incompatible materials are mixed or improperly used. All containers used for hazardous materials must be properly handled, and must be correctly labeled. Proper labeling includes the name of the manufacturer, address, hazards, safety instructions, and other precautions. Whenever chemicals are removed from their original, labeled containers, and are used for more than one work period the new container must also be labeled. Procedures should be established to prevent the return of unused materials to the wrong containers. Combining or substituting chemicals should be prohibited, except as provided in written procedures or as under the control of qualified personnel.

Internal waste treatment and licensed disposal facilities are approved disposal methods. Among physical characteristics to consider are toxic vapors or fumes, corrosiveness, and oxidizing properties of chemicals and hazardous materials.

Flushing to storm sewers in any quantity exceeding "de minimus" amounts is illegal for wastes classified as hazardous under the U.S. Resource Conservation and Recovery Act (RCRA). Check also with local, city,

and airport authorities for additional regulations that may prevail.

Spills

A spill or unintentional release of a toxic or hazardous substance into the environment may need to be reported to several agencies, depending on what was spilled, the quantity that was spilled, and other variables such as location, time, and effects of the spill itself. Several regulatory requirements may be in effect in an area. Management should research these regulations and prepare procedures accordingly, including a list of agencies that must be called under what circumstances. The form found in Figure 3-2 may be used to collect such information.

Three methods are used for cleaning up spills:

1. Flushing
2. Commercial absorbents or adsorbents
3. Fixants

If moderate to large quantities of a spilled material are involved, some of the material might be recovered for reuse. Check with local authorities having jurisdiction regarding regulations that may apply. Then the remainder can be cleaned up by means of one of the three methods. (The factors to consider in deciding whether recovery is advisable are given later in this section.)

1. Flushing

Small quantities of nonrecoverable residue can be "flushed" with excess amounts of diluting water. Flushing spills down the drain, however, is not always safe or appropriate. It definitely should not be attempted on large quantities of water-reactive materials because an explosion or fire could result. Before flushing is attempted, it is necessary to know that the material can be safely and legally cleaned up by this method.

2. Commercial absorbents or adsorbents

Small quantities of nonrecoverable residue can be cleaned up with commercial absorbent or adsorbent agents. Spill residues can be solidified with commercial batting materials, or with clays, kiln dust, fly ash, fuller's earth, soil, sand, or sawdust. Note that if the solidified waste is classified as hazardous, it cannot be legally dumped in a landfill; so, careless spills can result in very high costs to dispose of the residues.

3. Chemical fixants

Small quantities of nonrecoverable residue can be cleaned up with commercial fixant agents. Chemical fixation of spill residues can be achieved with certain commercial solidification products. Use of such products for minor spill residue cleanups may broaden the options for legal disposal and reduce costs.

The decision about whether to attempt spill recovery must take into account the potential risk of human exposure to hazardous materials during the recovery process. Factors that might eliminate the possibility of recovery would include proximity of a dangerous spill to a water supply or body of water, to a fire hazard, or to personnel or equipment.

These principles should be followed in spill handling programs:

- Plan methods that will allow material recovery to the maximum degree safely possible.
- Use cleanup agents that will not present disposal problems.
- Train employees in how to handle spills.
- In the United States monthly generation of more than 100 kg (45 lb) of spill cleanup residue and/or storage of more than 1,000 kg (453 lb) of such residue for more than 90 days is prohibited unless specifically permitted by the state regulatory agency for waste management or by the EPA under the Resource Conservation Recovery Act (RCRA).

Emergency information

Water dilution of most chemicals reduces the harmful effects of them on the human body. Eyewash fountains, emergency showers or water hoses also should be provided as required. Doctors and hospitals should have a list of hazardous chemicals and their antidotes so medical treatment, when necessary, can be given without delay. This can be supplemented by providing treating professional a copy of the relevant MSDS.

Those chemicals requiring the same specific fire extinguishing agent should all be stored together. The fire extinguisher should be conspicuously mounted immediately adjacent to the storage area. Knowledge of chemicals used and proper handling will reduce the possibility of fire.

Additional information on storage and handling is available in Part 1910 of Occupational and Safety Health Standards. (See References under U.S. Department of Labor; see also Chemical Manufacturers Association in References.)

Flammable solvent storage

Bulk flammable solvents should be stored out of doors, in protected areas, or in well-ventilated rooms with explosion-proof electrical services, including light fixtures, intercoms, telephone accessories, and ventilation equipment. Storage facilities for bulk flammable solvents should be physically separated from other storage and operations. Construction of such facilities should be in accord with the National Fire Codes published by the National Fire Protection Association. (See References.) Additional information and requirements on construction and storage can be found in Part 1910 of Occupational and Safety Health Standards (see References, under U.S. Department of Labor).

Because of flammability hazards, leaking drums should be removed immediately from enclosed spaces to prevent the buildup of hazardous vapor concentrations.

The preferred liquid dispensing method is by hand pumps listed by Underwriters Laboratories Inc. (UL). If spigots are used, they should be a listed, self-closing type. Drip pans to catch spilled liquids should be avoided unless the pans drain into a closed container. Safety containers, with the name of the solvent clearly printed on the container along with other required labeling, should be used on the job. When flammable liquids are being dispensed, bonding and grounding are required to prevent static generation and explosion.

HAZARDOUS MATERIAL SPILLAGE OR BREAKAGE REPORT

PLANE NO: _____ FLT. NO: _____ DATE _____

STATION STOPS: _____

PROPER NAME OF PRODUCT: _____

QUANTITY OF PRODUCT: PER UNIT: _____ UNITS PER PKS: _____

NO. OF PKGS CARRIED: _____

TYPE OF PRODUCT: LIQUID _____   GAS _____   SOLID _____

DUST _____   GRANULAR _____

EXACT LOCATION CARRIED ON AIRCRAFT: _____

NAME OF CONSIGNEE: _____ CONSIGNOR: _____

TYPE OF WARNING LABELS AFFIXED TO PACKAGE:

COLOR: _____WORDING: _____

DESCRIPTION OF PACKAGING: _____

_____

STATION DAMAGE OR SPILLAGE FOUND: _____ TIME AND DATE: _____

DESCRIPTION OF DAMAGE: _____

_____

NUMBER OF PERSONS EXPOSED TO PRODUCT: _____

WHO REPORTED INCIDENT: _____

WHAT GOVERNMENTAL AGENCIES HAVE BEEN NOTIFIED:

	NAME OF PERSON	PHONE NO.	TIME AND DATE
FAA:			
NRC:			
AIRPORT AUTHORITY:			
USPHS:			
OTHER:			

REPORT COMPLETED BY: _____ DATE _____

Figure 3-2. When hazardous materials are spilled or their containers are broken, a checklist form such as this one should be filled out to alert company safety personnel and the appropriate government agencies.

Spills should be wiped up immediately. Rags or absorbent compound used in the cleanup should be placed in listed waste containers. These containers should be disposed of only in a manner approved by the government authority having jurisdiction.

Housekeeping is a prime concern where flammable and volatile solvents are used. Hence, housekeeping measures should be regularly reviewed and upgrading should take place whenever indications are that housekeeping methods have deteriorated.

If possible, operations with volatile solvents should be isolated. Prolonged exposure to levels above the allowable limit should be avoided through methods discussed earlier and through administrative controls. Operators are to wear appropriate personal protective equipment when necessary.

If spills occur operators should remove the exposed clothing quickly, and thoroughly wash the exposed skin. Contaminated clothing should be laundered or discarded, depending on the contaminant material. Operators handling solvents should wash their hands before eating or drinking. Smoking should not be allowed around volatile chemicals as a matter of course. (For additional information, see, in References, Part 1910 of Occupational and Safety Health Standards listed under U.S. Department of Labor.)

Flammable solvent disposal

Flammable wastes should not be dumped into public sewage or drainage systems. In case of incidental spills into public sewage or drainage systems, copious amounts of water should be used to flush the system. The appropriate governing authority and the local department should be notified.

Solid flammable materials contaminated by flammable liquids should be placed in covered metal containers designed for this purpose and removed from work areas daily. Waste flammable liquids may also be saved and later burned in firefighting training exercises by the airport's private fire department if permitted by local regulations.

In any event, wastes should be disposed of properly. If not by burning in accordance with local regulations, wastes should be disposed of by a reputable EPA-registered contractor. Wastes awaiting disposal should be stored properly.

Compressed gases

The following gases are used in aviation:

- **Acetylene, hydrogen, argon** and **helium,** used for welding
- **Oxygen,** for breathing and welding
- **Nitrogen,** for tire and strut inflation and as a fire-extinguishing propellant

Other agents include:

- **Carbon dioxide,** a fire-extinguishing agent
- **Refrigerant 12** (dichlorodifluoromethane), a refrigerant
- **Refrigerant 1301** or **Halon** (bromotrifluoromethane), a fire-extinguishing agent
- **Compressed air,** for emergency equipment on aircraft and many other uses
- **Liquefied petroleum gas** (butane and propane), a fuel used for motor vehicles, torches, burners and heaters

Pressurized cylinders may contain pressures up to 41,500 kPa (6,000 psi). A direct relationship exists between temperature and pressure. Most cylinders are designed and constructed according to U.S. Department of Transportation regulations and should bear the U.S. Department of Transportation, American Society of Mechanical Engineers (ASME) or equivalent cylinder stampings, found near the neck of the cylinder. Some cylinders have reinspection and hydrostatic testing dates stamped in the same place.

Cylinder contents should be legibly marked on each cylinder.

Serious incidents with compressed gases can generally be traced to misuse, abuse, or mishandling of the gas, cylinder, or related equipment.

Inside cylinder storage areas should be clean and free from excessive heat. Flammable gases should be stored separately from other gases, and oxygen should be stored separately from flammable gases.

Storage areas should have effective natural ventilation or listed mechanical ventilation appropriate to the exposure. The buoyancy of the gas in storage areas helps determine the proper ventilation. Gases lighter than air should be exhausted at the top while gases heavier than air should be exhausted at floor level.

In storage areas, cylinders of one type should not be mixed with other types of cylinders. Floors should be level and constructed of noncombustible materials.

Outside storage requires all the protection afforded by the inside storage facility, plus a non-combustible canopy to protect cylinders from weather.

Cylinders should stand upright, with the valve up, especially those of acetylene. Empty cylinders should be stored apart from full cylinders and the word "empty" or the letters "MT" chalked in large letters on them. Valves should be closed and protective caps installed. Storage areas should be free of ignition sources, including smoking materials.

The authority having jurisdiction should be consulted about flammable gas storage. The storage area should be provided with chains (or similar devices) to keep cylinders from being knocked over. Areas near elevators, stairs, ramps, or busily traveled aisles/locations should not be used for storage. Storage of cylinders should facilitate their being used in the order which they are received.

Newly delivered cylinders should be checked for DOT or ASME stampings, condition, valve, and content labeling.

All compressed gas cylinders should be visually inspected for defects such as excessive rusting or dents. Cylinders with rust, dents, or missing protective caps should not be accepted from the supplier.

Valves or safety devices must not be tampered with. A leaking cylinder should be taken outside the building and tagged as defective. Leaking cylinders should never be accepted from a supplier.

Because of their shape and weight, most cylinders are difficult to handle manually. However, if such handling is necessary, cylinders should be tipped slightly and rolled on the bottom edge – not dragged or slid across the floor. Mechanical handling of cylinders generally requires carts or specially constructed skids. Using electromagnets or slings is not acceptable for mechanical handling. Bars should not be used, because inserting bars through the protective caps may damage valve assembly.

Cylinders should not be dropped or permitted to strike each other. They should not be used as rollers or for any purpose other than gas containers. A direct flame or arc never must be permitted to contact a cylinder.

Before compressed gas cylinders are connected to a system valve, outlets on all cylinders should be cleared – except for cylinders containing hydrogen or toxic gases. The valve should be pointed away from the body and away from other people, then opened slightly to release foreign material.

The compressed gas cylinder never should be used without a pressure-reducing regulator attached to the valve or if cylinders are in a manifold header. There are regulators and pressure gauges designed for use with specific types of gas. Only those regulators and pressure gauges designed for the specific gas should be used. No one should attempt to force connections that do not fit easily. A Compressed Gas Association standard specifies various size threads for cylinders containing different gases; some have left-hand threads to frustrate any attempt to use a wrong regulator (see References). If a leak develops between the cylinder and regulator, the gas should be shut off at the bottle before attempting to tighten the union.

All compressed gas valves should be opened slowly to prevent sudden surges of pressure that may damage regulators. An oxygen valve, for example, should be opened slowly to prevent surge and possible fire, then opened fully to gain the sealing benefit of the double-seated valve. Valves on acetylene cylinders should be opened approximately 1½ turns. The special tool recommended by the supplier should be the only tool used for these valves. The tool is to be left on the valve so that the gas can be shut off quickly in an emergency. Compressed gas cylinder valves should be closed when gas is not being withdrawn.

When high- and low-pressure taps are used on the same cylinder, hoses should be different colors. Fittings on the hose ends should be a different type and size.

Oil and grease never should be used as a lubricant for oxygen cylinders or appurtenances because an explosion may result. Oxygen cylinders and fittings should be kept away from oil and grease, and such cylinders or apparatus should not be handled with oily hands, gloves, or clothing. Operators should wash their hands thoroughly before handling oxygen cylinders or systems. Oxygen should never be used as a substitute for compressed air in pneumatic tools or for any unapproved purpose.

Cylinders must not be taken into tanks or other unventilated areas.

Manifolds for compressed gas cylinders should be a type approved by the gas supplier or the appropriate standards and specification group, such as the Compressed Gas Association or the National Fire Protection Association.

Connections should be tested for tightness with conventional leak testing products. Testing for leaks by an open flame is prohibited.

Bottled oxygen and nitrogen are produced under rigid quality controls; however, as cylinders are rotated from one industry to another, contamination may result. Although such occurrences are rare, consequences are severe. If there is any suspicion of contamination, contact the supplier of the cylinder.

OTHER HAZARDOUS MATERIALS AND HARMFUL PHYSICAL AGENTS

Aviation employees may be exposed to other hazardous materials and processes, such as noise, poor lighting conditions, radiation and explosive devices used to activate aircraft fire protection systems.

Noise

Noise can affect employees in several ways:

- Hearing loss
- Shift in hearing threshold
- Fatigue and irritability
- Communications interference
- Increased blood pressure

Noise exposures are prevalent throughout the aircraft industry. Potentially damaging levels are common. All companies should install a hearing conservation program. The program should include at least the following:

- Sound level measurements to determine which areas exceed acceptable limits
- Medical surveillance – preplacement, periodic, and postemployment – by audiometric examination
- Noise control/reduction programs
- Administrative controls for employees in noise areas
- Personal hearing protection
- Employee training and education

Noise control principles and techniques are varied and sophisticated. (See Chapter 13, "Personal Protective Equipment and Guarding," for more information on noise control. Also see, in References for this chapter, National Safety Council *Noise Control Manual,* American Industrial Hygiene Association *Noise and Hearing Conservation Manual,* NIOSH *Industrial Noise Control Manual* under U.S. Department of Health and Human Resources, and the American Petroleum Institute "Guidelines on Noise.")

For severe noise control problems, a professional consultant may be engaged.

Nonionizing radiation

Nonionizing radiation refers to that portion of the electromagnetic spectrum where energies of emitted photons are insufficient to disrupt absorbing molecules. Included in this portion of the spectrum are microwave energy, ultraviolet, visible and infrared radiation. Sources of microwave radiation in the aviation industry include radar, telecommunications and high frequency heating. Ultraviolet, visible and infrared sources include welding, plasma spray, lasers and high-temperature heated sources.

Depending on wavelength, the effects of nonionizing radiation (except for ordinary visible light) can be harmful: Microwave energy can cause thermal damage to eyes, testes and the central nervous system; lasers can cause damage to the retina; infrared can cause heat stress; and ultraviolet can result in photochemical effects to the eye and skin, skin cancer, and photosensitivity. (See Michaelson, 1974, in References for a review of nonionizing radiation health problems, and "Nonionizing Radiation," Chapter 11 in the National Safety Council's *Fundamentals of Industrial Hygiene*.)

Microwave ovens. Microwave oven use is well established. Depending upon their age and amount of use, ovens can leak microwave energy in excess of the 5 milliwatts per square centimeter (5 mW/cm2) allowed as noted in ANSI C95.1. (See References.) It is good practice to check ovens periodically for leakage. Commercial instruments are available for this purpose. After a defective oven has been removed from service and repaired, another check for microwave leakage is recommended.

Lighting. Poor lighting (glare, low or uneven light intensity) can cause eyestrain or low efficiency. (See the Illuminating Engineering Society's *Lighting Handbook,* listed in References, for detailed information on recommended lighting levels, lighting parameters and equipment specifications.)

Ionizing radiation

Some nondestructive testing operations use radioactive isotopes and X-radiographic equipment to examine critical aircraft parts. If the radioactive source is large enough, these operations must conform to requirements of the U.S. Nuclear Regulatory Commission (NRC) and/or other government regulations. A qualified safety officer should be assigned responsibility for radioactive materials, and the handling and monitoring of X-ray-producing equipment.

Components like electron tubes, spark gaps, and luminescent signs contain radioactive material. The NRC and other authorities having jurisdiction have established standards for the handling, storage, use and disposition of radioactive materials. Local procedures should conform to these standards.

The quantity of radioactive materials contained in units like those previously mentioned is such that no significant external radiation hazard exists when handled individually or in small numbers. The most severe hazard would be encountered if one or more units broke down. For this reason, units should be stored in their packing containers. Units should remain in these containers until used as replacements, and the replaced unit should be returned to the container. No radioactive component should be disassembled or discarded at random. Maintenance procedures must include instructions for returning units that need repair or maintenance to the supplier or distributor. Disposition of unserviceable units as well as those broken down should conform to government regulations.

Shipments of radioactive materials in aircraft present another potential source of exposure to ionizing radiation if breakage and spillage occur. U.S. Federal Aviation Administration and Department of Transportation regulations as detailed in Dangerous Goods Regulations must be complied with (see References).

Explosive devices

Presently there are two types of explosive devices used to activate aircraft fire protection systems. One device, commonly called an "initiator," is used to release the extinguishing system in the fuel tank's vents or wing surge tanks. These initiators resemble blasting caps and, when not connected to an extinguisher, are subject to handling and storage procedures compatible with all laws and ordinances applicable to explosives and blasting devices.

Initiators must be packaged individually and the package certified to sustain and contain detonation of one initiator within an explosive atmosphere. All packages should be sealed with wires and/or lead seals or similar devices.

Initiators used or unused should remain packaged in the portable magazine while being stored and transported intraplant and interfield. Magazines should be painted red, the standard explosive label should be affixed to them and the shelf life date marked on the container's exterior surface.

Provisions should be made to ensure safe disposition of units whose shelf life has expired. Shelf life should mean the period from date stamped on the unit or its nameplate.

Another explosive device is called a "squib." This unit has the force of a .45-caliber shell and is used for activating the engine fire-extinguishing system. Their shipping is governed by International Air Transport Association Regulation #1687, Note 200. (See References.) The units are packaged in quantities of one each for storage and issue purposes. When the shelf life of these units has expired, safe disposition is required. Often, local police authorities will accept these units for safe disposition. They should not be disposed of with local trash.

REFERENCES

American Conference of Governmental Industrial Hygienists:
Industrial Ventilation.
Air Sampling Instruments.
American Industrial Hygiene Association:
Noise and Hearing Conservation Manual, 4th ed., 1986.
Hygienic Guide Series.
American National Standards Institute:
Safety Levels with Respect to Human Exposure to Radio Frequency Electromagnetic Fields C95.1-1991 (R1997).
Installation Using Non-Medical X-ray and Sealed Gamma-Ray Sources, Energies Up to MeV, General Safety Standards for, ANSI/NBS 114.
Practices for Respiratory Protection, ANSI Z88.2-1980.
American Petroleum Institute. *Guidelines on Noise, Medical Research Report EA-7301.*
Chemical Manufacturers Association
Compressed Gas Association
International Air Transport Association. *Dangerous Goods Regulations,* published annually.
Hamilton and Hardy, *Industrial Toxicology,* Publishing Sciences Group Inc., 1984.
Illuminating Engineering Society of North America. *IES Lighting Handbook.*
Michaelson, Sol, "Standards for Protection of Personnel against Nonionizing Radiation," *American Industrial Hygiene Association Journal,* 35:766 (1974).
National Fire Protection Association:
System for the Identification of the Hazardous Materials for Emergency Response, NFPA 704, 1996
Flammable and Combustible Liquids Code, NFPA 30, 1996
National Loss Control Service Corp. *Right to Know Compliance Manual.*
National Safety Council. *Fundamentals of Industrial Hygiene,* 4th ed. 1996.
Patty's Industrial Hygiene and Toxicology: Toxicology, vol. 2A, 1993; vol. 2B, 1994; vol. 2C, 1994. John Wiley and Sons.
Sax, Irving N., *Dangerous Properties of Industrial Materials,* 9th ed., Van Nostrand Reinhold, 1996.
Technical Operations, Inc., Radiation Products Division. *Isotope Radiography, Radiation Safety Handbook.*
U.S. Department of Health and Human Resources, National Institute for Occupational Safety and Health (NIOSH):
Industrial Noise Control Manual.
NIOSH-Certified Personal Protective Equipment.
U.S. Environmental Protection Agency:
The Effects on Populations of Exposure to Low Levels of Ionizing Radiation (BIER Report). Published by Office of Radiation Programs. Available from Criterion Standards Division AW-460.
Toxic Substances Control Act.
U.S. Government Printing Office. *Suspected Carcinogens – Subfile of the Toxic Substances List.*
U.S. Occupational Safety and Health Administration (OSHA). *Occupational Safety and Health Standards in Code of Federal Regulations,* Title 29, Part 1910.

ACCIDENT INVESTIGATIONS

CHAPTER 4

Thorough accident investigation is crucial to a good safety program because it provides the necessary information for preventing similar accidents.

NOTE: IN ANY DISCUSSION OF ACCIDENT INVESTIGATIONS, IT IS NECESSARY TO DEFINE THE WORD "ACCIDENT." AN ACCIDENT IS ANY UNEXPECTED EVENT THAT INTERRUPTS AN OPERATION. ACCIDENTS DO NOT NECESSARILY RESULT IN INJURY OR DAMAGE; THEY ARE ALWAYS PRECEDED BY INADEQUATE PROCEDURES, UNSAFE CONDITIONS OR BOTH.

THE PURPOSE OF AN ACCIDENT INVESTIGATION IS TO DETERMINE THE CAUSES OF AN ACCIDENT, DEVELOP CORRECTIVE MEASURES, AND MAKE RECOMMENDATIONS FOR ELIMINATION OF THE CAUSES.

Gathering information

One question often asked is, "Which accidents should be investigated?" Because all accidents are potentially serious, all should be investigated. Here are some other basic guidelines for accident investigation.

Who should conduct the investigation?

The person or persons investigating the accident should have some knowledge of all the circumstances surrounding the accident: the work process, the equipment, the employee involved in the accident, and accident sources and causes. The person who most completely meets this criterion is usually the immediate supervisor of the employee involved in the accident.

The company safety leader, the safety committee or an insurance company representative also may assist in the investigation – or perform a supplementary investigation.

When should the investigation be conducted?

The investigation should be conducted as soon as possible. Delays – even of only a few hours – can permit important information or items to be removed, destroyed or forgotten. If possible, every employee who is involved in, or is a witness to, the accident should be questioned. It is important that employees being questioned know that the purpose of the investigation is not to find fault or blame – it is to determine the root causes in order to prevent similar accidents.

The investigator should try to avoid jumping to conclusions: It is very important that he or she uncover the underlying or root causes of the accident. The investigator must ascertain why an unsafe condition/procedure existed or why the unsafe behavior occurred. Simply finding or identifying unsafe conditions is not enough. An apparently reasonable conclusion about the cause of an accident will often be changed when facts that did not first appear to be important are explored.

While gathering accident facts, the investigator should encourage employees to give their ideas on how the accident could have been prevented.

Preparing an accident report

Once the investigation of an accident has been performed, a report should be prepared that describes the findings in detail. The report should answer all questions pertinent to the accident and list the corrective actions to prevent a recurrence. The person preparing the report should follow these rules:

• Stick to the facts.
• Weigh the value of the facts.
• Reach justified conclusions.

The questions in the following section should be a helpful guideline for writing the report:

Who was involved? Accidents usually involve more than just the injured person. Very often more than just a single person contributed to the cause of an accident. Therefore, the investigation should go beyond just the person or persons directly involved. It also should be determined who directed the employee to perform the particular job function, who trained the employee, and who failed to report an unsafe condition or procedure.

Where did the accident occur? Again, the investigator must look beyond the obvious. The name of the department is not enough. A detailed description of the accident site should be included. Also, the investigator should determine if the people involved were supposed to be where they were. Was the equipment in its proper location?

What happened? This question can be broken down further to uncover the following facts:

1. *What was being done?* The answer describes an action or procedure.
2. *What things were involved?* The answer could be a description of the tool, material or equipment that was involved.
3. *What was the result?* This is answered by describing the actual injury, including the nature of injury and the part of the body injured, and/or detailing the equipment and property damage.

When did the accident occur? The answer to this question requires more than just recording the date. The time of day and the day of the week also can be very important. Other questions that may be asked are: Did the accident occur at the beginning or end of a shift? Was it early or late in the day, and when did the employee start working? What were the light conditions or weather conditions (if applicable) at the time of the accident?

How did the accident happen? The answer to this question brings together all the facts of the accident. The answer is a description of the people, things, places, and time as they all combine into a complete event. The exact sequence of events that led to the accident should be reported. The person preparing the report should write the step-by-step sequence of how the accident happened; the list should include everything that occurred.

Why did the accident occur? In order to determine or recommend what corrective action should be taken, it must be determined exactly why the accident occurred.

Do not always settle for the first cause identified. To identify root causes, be sure to ask follow-up questions such as:

- Why was the injured person inattentive?
- Why was he or she poorly trained?
- Why did someone fail to report an unsafe condition or procedure?
- Why did what happened produce an accident?
- Why did the combination of all the factors that made up the event result in an injury?
- Why did the event result in anything other than an ordinary, everyday occurrence? Were there any other conditions that contributed to the accident?

Recommended corrective action. After evaluating the facts, the investigator most likely will find the accident was caused by a combination of unsafe conditions and inadequate job procedures. Recommendations to prevent a recurrence should pinpoint the unsafe condition or inadequate procedure, and should be designed to correct all contributing factors.

The report should now be submitted to top management. After the report has been submitted, it is very important to follow up to make sure that recommended changes are implemented correctly.

Here is a summary of the important steps in preparing an accident report: Remember, all accidents should be investigated as soon after the accident as possible. All people involved should be interviewed in an effort to determine exactly how the accident occurred. Once all the facts have been put together, a report should be submitted to top management. The report should answer the questions *who, where, when, what, how* and *why*. Also, no investigation is complete unless corrective action is suggested.

ACCIDENT RECORDS

The form of an accident investigation report may vary, but any report should contain all the essential information. This section contains examples of an accident report form (see Figure 4-1).

A well-designed safety program seeks to know not only the causes of accidents, but also their costs. From accident investigation reports and other company sources, a report can be produced that includes the following:

• Number of accidents by causes
• Cost of accidents by causes
• Accident causes by department and/or location
• Accident costs by department and/or location
• Accident summaries by shift

Figure 4-1 shows an example of a report form that can be used to aid in safety efforts. Often, the insurance carrier can assist in making accident investigation reports. Accident information that is going to be computerized can be coded for ease of handling. Later in this section, examples are given of codings used by companies in producing reports.

Records on accidents also can be used to:

• Create interest in safety among members of management, supervisors, and safety committees by furnishing them with information about the accident experience of their departments.
• Help determine the principal sources of accidents so efforts can be focused in directions that will effect the largest reductions in accidents.
• Judge safety program effectiveness by showing whether accident experience is improving or getting worse.
• Make possible comparisons with other companies, industries or groups.

INVESTIGATION REPORT

The preceding section described general principles and procedures that apply to any accident investigation. This section tells how to complete the Accident Investigation Report, a form that serves as an investigation guide and as a record of the facts obtained (Figure 4-1). The report is one of three key documents. The second key document, the Guide for Identifying Causal Factors and Corrective Actions (Figure 4-3), is described in the next section. The third key document, the Summary of Causal Factors (Figure 4-4), helps investigators compile and analyze data from many accidents.

When used together, the report and the guide tell investigators what questions to ask, what factors to investigate and what other information to document as part of the permanent record of the accident and its aftermath. They also help investigators identify causal factors and corrective actions.

The report is designed primarily for investigation of accidents involving injuries, but also can be used to investigate occupational illnesses arising from a single exposure (for example, dermatitis caused by a splashed solvent or a respiratory condition caused by the release of a toxic gas).

All questions on this form should be answered. If no answer is available, or the question does not apply, the investigator should so indicate. Answers should be complete and specific.

Supplementary sheets can be used for other information, such as drawings and sketches, and should be attached to the report. A separate form should be completed for each employee who is injured in a multiple-injury accident.

The individual entries are explained below.

Department: Enter the department or other local identification of the work area to which the injured is assigned (for example, maintenance shop or shipping room). In some cases, this may not be the area in which the accident occurred.

ACCIDENT INVESTIGATION REPORT

Case Number

Company	Address
Department	Location

1. Name of injured	2. Social Security Number	3. Sex	4. Age	5. Date of accident ___ / ___ / ___

6. Home Address	7. Employee's usual occupation	8. Occupation at the time of the accident

9. Employment category
☐ Regular, full-time ☐ Regular, part-time ☐ Temporary
☐ Seasonal ☐ Non-Employee

10. Length of employment
☐ Less than 1 mo. ☐ 1-5 mos.
☐ 6 mos. - 5 yrs. ☐ More than 5 yrs.

11. Time in occup. at time of the accident
☐ Less than 1 mo. ☐ 1-5 mos.
☐ 6 mos. - 5 yrs. ☐ More than 5 yrs.

12. Nature of injury and part of the body

13. Case numbers and names of others injured in same accident

14. Name and address of physician

15. Name and address of hospital

16. Time of injury
A.   a.m.
           p.m.

B. Time within shift _____

C. Type of shift _____

17. Severity of injury
☐ Fatality
☐ Lost workdays - days away from work
☐ Lost workdays - days of restricted activity
☐ Medical treatment
☐ First Aid
☐ Other, specify _____

18. Specific location of accident

On employer's premises?   ☐ Yes   ☐ No

19. Phase of employee's workday at time of injury
☐ During rest period   ☐ Entering or leaving plant
☐ During meal period   ☐ Performing work duties
☐ Working overtime   ☐ Other

20. Describe how the accident occurred

21. Accident sequence. Describe in reverse order the occurence of events preceding the injury and accident.
Start with the injury and moving backward in time, reconstruct the sequence of events that led to the injury.

A. Injury event _____

B. Accident event _____

C. Preceding event #1 _____

D. Preceding event #2, 3, etc. _____

22. Task and activity at time of accident
General type of task _____
Specific activity _____
Employee was working:
☐ Alone
☐ With crew or fellow worker
☐ Other, specify

23. Posture of employee

24. Supervision at time of accident
☐ Directly supervised ☐ Indirectly supervised
☐ Not supervised   ☐ Supervision not feasible

Accident Investigation Report   Page 1

Figure 4-1. This form asks for the circumstances contributing to the incident, as well as for a recommendation for avoiding similar incidents.

25. Causal factors. Events and conditions that contributed to the accident.
Include those identified by use of the **Guide for Identifying Causal Factors and Corrective Actions.**

26. Corrective Actions. Those that have been, or will be, taken to prevent recurrence.
Include those identified by use of the **Guide for Identifying Causal Factors and Corrective Actions.**

Prepared by _____

Title _____

Department _____ Date _____

Approved _____

Title _____ Date _____

Approved _____

Title _____ Date _____

Developed by the National Safety Council
©2006 National Safety Council

Location: Enter the location where the accident occurred if different from the employer's mailing address.

Accident investigation report

1. Name of injured: Record the last name, first name and middle initial.

2. Social Security Number

3. Sex

4. Age: Record the age of the injured at his/her last birthday, not the date of birth.

5. Date of accident or initial diagnosis of illness

6. Home address

7. Employee's usual occupation: Give the occupation to which the employee is normally assigned (for example, assembler, lathe operator or clerk).

8. Occupation at time of accident: Indicate the occupation in which the injured was working at the time of the accident. In some cases, this may not be the employee's usual occupation.

9. Length of employment: Check the appropriate box to indicate how long the employee has worked for the organization.

10. Time in occupation at time of accident: Record the total time the employee has worked in the occupation indicated in item 8.

11. Employment category: Indicate injured's

employment category at the time of the accident (for example, regular, temporary or seasonal).

12. Case numbers and names of others injured in same accident: For reference purposes, the names and case numbers of all others injured in the same accident should be recorded here.

13. Nature of injury and part of body: Describe exactly the kind of injury or injuries resulting from the accident and the part or parts of the body affected. For an occupational illness, give the diagnosis and the body part or parts affected.

14. Name and address of physician

15. Name and address of hospital

16. Time of injury: In part B, indicate in which hour of the shift the injury occurred (for example, first hour). In part C, record the type of shift (for example, rotating or straight day).

17. Severity of injury: Check the highest degree of severity of injury. The options are listed in decreasing order of severity.

18. Specific location of accident: Indicate whether the accident or exposure occurred on the employer's premises. Then record the exact location of the accident (for example, at the feed end of #2 assembly line or in the locker room). Attach a diagram or map if it would help identify the location.

19. Phase of employee's workday at time of injury: Indicate what phase of the workday the employee was in when the accident occurred. If "other," be specific.

20. Describe how the accident occurred: Provide a complete, specific description of what happened. Tell what the injured and others involved in the accident were doing prior to the accident; what relevant events preceded the accident; what objects or substances were involved; how the injury occurred; the specific object or substance that inflicted the injury;

and what, if anything, happened after the accident. Include only facts obtained in the investigation. Do not record opinions or place blame.

21. Accident sequence: Provide a breakdown of the sequence of events leading to the injury. This breakdown enables the investigator to identify additional areas where corrective action can be taken.

In most accidents the accident event and the injury event are different. For example, suppose a bursting steam line burns an employee's hands or a chip of metal strikes an employee's face during a grinding operation. In these cases the accident event (the bursting of the steam line or the setting in motion of the metal chip) is separate from the injury event (the steam burning the employee's hands or the chip cutting the employee's face). The question is designed to draw out this distinction and to record other events that led to the accident event.

Although not accident events themselves, events preceding the accident event may have contributed to the accident. These preceding events can take one of two forms. They can be something that happened that should not have happened, or something that did not happen that should have happened. The steam line, for example, may have burst because of excess pressure in the line (preceding event #1). The pressure relief valve may have been corroded shut, preventing the safe release of the excess pressure (preceding event #2). The corrosion may not have been discovered and corrected because a regular inspection and test of the valve was not carried out (preceding event #3).

To determine whether a preceding event should be included in the accident sequence, the investigator should ask whether its occurrence (if it should not have happened) or nonoccurrence (if it should have happened) permitted the sequence of events to continue through the accident event and injury event.

Take enough time to think through the sequence of events leading to the injury and to record them separately on the report. This information can be used in the Guide for Identifying Causal Factors and Corrective Actions (Figure 4-3) to help identify management system defects that contributed to the events in the accident sequence. By identifying such defects, management can help prevent many more types of accidents than just the type under investigation.

In the example, the failure to detect the faulty pressure relief valve should lead to a review of all equipment inspection procedures. This review could prevent other accidents that might have resulted from failure to detect faulty equipment in the inspection process.

Additional sheets may be needed to list all events involved in the accident.

22. Task and activity at time of accident: In parts A and B, first record the general type of task the employee was performing when the accident occurred (for example, pipe fitting, lathe maintenance or operating punch press). Then record the specific activity in which the employee was engaged when the accident occurred (for example, oiling shaft, bolting pipe flanges or removing material from press). In part C, check the appropriate box to indicate whether the injured employee was working alone, or with a crew or co-operator.

23. Posture of employee: Record the injured's posture in relation to the surroundings at the time of the accident (for example, standing on a ladder, squatting under a conveyor or standing at a machine).

24. Supervision at time of accident: Indicate in the appropriate box whether, at the time of the accident, the injured employee was directly supervised, indirectly supervised or not supervised. If appropriate, indicate whether supervision was not feasible at the time.

25. Causal factors: Record the causal factors (events and conditions that contributed to the accident) that were identified by use of the Guide for Identifying Causal Factors and Corrective Actions (Figure 4-3).

26. Corrective actions: Describe the corrective

actions taken immediately after the accident to prevent a recurrence, including the temporary or interim actions (e.g, removed oil from floor) and the permanent actions (e.g., repaired leaking oil line). Record other recommended corrective actions selected from the Guide for Identifying Causal Factors and Corrective Actions. Include any other corrective actions as requested.

NOTE: USERS MAY ADD OTHER DATA ELEMENTS TO THE FORM TO FULFILL LOCAL OR CORPORATE REQUIREMENTS. TYPES OF DATA ELEMENTS THAT MIGHT BE ADDED INCLUDE:

- Information on accident patterns that is specific to a particular industry or organization (e.g., an establishment with many confined-space accidents might wish to add some questions on accidents of that type).

- Information required for special studies (for example, a study tracing the effectiveness of a specific corrective action).

- More detailed severity information (e.g., the cost of the accident).

- Management data for use in performance reviews and in determining training needs.

- Exposure data for use in calculating incidence rates for injuries associated with certain activities. This data would be estimates or the actual number (or percent) of hours that the employees devote to the activity in a week, month, or year. Incidence rates based on the hours of exposure to specific activities can then be calculated. The use of incidence rates yields a fairer comparison of activities than methods comparing only the total number of cases associated with each activity.

GUIDE FOR IDENTIFYING CAUSAL FACTORS AND CORRECTIVE ACTIONS

The Guide for Identifying Causal Factors and Corrective Actions is the second key document in this chapter. It focuses on the four elements of a basic system: equipment, environment, people and management (Figure 4-2). These four system elements are combined to make products and profits. But sometimes they work together in unexpected ways to produce accidents.

The accident investigation method should focus on three key tasks. The first is identification of the causal factors that resulted in the accident. The second is identification of the corrective actions that will minimize the likelihood of a similar accident and also will minimize the severity or adverse consequences of a similar accident if it should occur. The third task is selection of the corrective actions that have the best chance of reducing the risk.

Identifying causal factors

The causal factor identification procedure in the guide can be applied to any workplace accident. It is based on a simple "Yes" or "No" response to a series of questions. The Guide is divided into four parts: Equipment, Environment, People and Management. Each part has one or two basic questions. Answers to the basic questions determine how to proceed through the other questions in that part. For example, the first

basic question in Part 1, Equipment, is "Was a hazardous condition(s) a contributing factor?" If the answer is "Yes," answer the remaining questions under "Causal Factors." If the answer is "No," proceed to the next part of the guide.

Answer all questions by placing an X in the "Y" circle or box for "Yes" or in the "N" circle or box for "No." Marks in the boxes will signify that the corresponding items are not causal factors. Marks in the circles will signify that the items are causal factors. Because accidents rarely have a single cause, this process usually yields more than one causal factor for each accident.

Use the "Comment" column to the right of the "Causal Factors" questions to record specifics about the accident under investigation. Provide a comment for each item identified as a causal factor; that is, items for which Xs appear in circles.

Identifying corrective actions

A corrective action is a response to eliminate a deficiency of some kind. The guide lists several "Possible Corrective Actions" for each causal factor. These suggested actions are stated in general terms. They guide the investigators in identifying specific corrective actions that relate to the accident they are investigating.

The "Possible Corrective Actions" are intended to strengthen the overall safety programs and eliminate or minimize such management system defects as oversights, omissions or lack of control. Of course, the list of actions in the guide cannot cover every imaginable contingency. Investigators, therefore, also should consider other corrective actions.

In the "Recommended Corrective Actions" column, list specific corrective actions that can be taken to minimize or eliminate the causal factors that resulted in the accident. The list should include remedies to eliminate defects that have been identified in the management system. Each recommended corrective action should be considered a candidate for implementation.

MANAGEMENT

PEOPLE

MISSION

EQUIPMENT   ENVIRONMENT

Figure 4-2. A basic system in which people, equipment, and environment are managed to accomplish a mission.

Selecting corrective actions

Most investigations will suggest several recommended corrective actions. Two or more corrective actions from this list of candidates often will be chosen for implementation. Some are bound to be more effective than others. And some are bound to be more costly than others. Factors that usually influence the selection include:

- Effectiveness
- Cost
- Feasibility
- Effect on productivity
- Time required to implement
- Extent of supervision required
- Acceptance by employees
- Acceptance by management

Corrective actions that best fulfill these criteria offer optimum possibilities for reducing risk.

SUMMARY OF CAUSAL FACTORS

The purpose of summarizing causal factors is to identify those that have contributed most frequently to a group of accidents being analyzed. When causal factors are found repeatedly in a number of accident investigations, they generally reveal patterns that suggest changes in the management system or in the safety and health program. Repeated analysis over months or years will show long-term trends. These also can be used to evaluate the impact of changes in the management system and the impact of corrective actions taken.

The third key document presented in this chapter, Summary of Causal Factors (Figure 4-4), suggests a way to summarize causal factors. Each statement in the summary corresponds to a question in the Guide for Identifying Causal Factors and Corrective Actions. The six columns of boxes next to the statements are provided for tallies of the frequency of occurrence of the causal factors. Essentially a tabulation form, the guide facilitates the tallying of causal factors by whatever general category and subcategories the analyst selects.

A two-step procedure can help identify causal factors in a given category. First the general category and subcategories are selected, and then the number of cases is recorded on the summary. Safety personnel usually carry out this kind of analysis.

Selecting categories

Many different categories can be selected for analysis. For example, an entire plant or establishment can be treated as a single category. The analysis would then tell management which causal factors were identified most frequently in the accident cases that occurred throughout the establishment in a given period. This information could suggest priorities for general changes in the management system to improve the overall safety and health performance.

Often it is useful to analyze the causal factors for subcategories within a general category. For example, if the entire plant is the general category, then departments or occupations could be the subcategories. Or if the occupation "assembler" is the general category, then the various amounts of time in the occupation could be the subcategories. If total accidents in a specific department is the general category, subcategories can deal with such factors as occupation or job category, age of worker, number of injuries, or parts of body injured.

Many of the items included in the Accident Investigation Report form can be selected as categories for analysis. The general category selected should be written on the summary form on the line provided, and all of the cases that fall into this category should be tallied in the column labeled "Category Total." If subcategories are used, they should be entered in the other columns. (Add more columns if necessary.) Cases that fall into the general category should be tallied in the "Category Total" column just as before, but the cases that also fall into one of the subcategories should be tallied again in the appropriate subcategory column.

One advantage of the summary is its flexibility. The analyst is encouraged to select any general category and any set of subcategories – they all will be related to causal factors.

REFERENCES

National Safety Council
    Industrial Data Sheet: Work Accident Records and Analysis, 527.
    Accident Investigation, 2nd ed. 1995.
    Accident Prevention Manual for Business and Industry, Administration & Programs volume, 11th ed. 1997.
U.S. Occupational Safety and Health Administration (OSHA).
    Recordkeeping Requirements under the Occupational Safety and Health Act of 1970

GUIDE FOR IDENTIFYING CAUSAL FACTORS & CORRECTIVE ACTIONS

Case Number

Answer questions by placing an X in the **"Y"** circle or box for yes or in the **"N"** circle or box for no.

PART 1 EQUIPMENT

O □
Y N

1.0 WAS A HAZARDOUS CONDITION(S) A CONTRIBUTING FACTOR?
If yes, answer the following. If no, proceed to Part 2.

	Causal Factors	Comment	Possible Corrective Actions	Recommended Corrective Actions
O □ Y N	1.1 Did any defect(s) in equipment/tool(s)/material to hazardous condition(s)?		Review procedure for inspecting, reporting, maintaining, repairing, replacing, or recalling defective equipment/tool(s)/material used.	
□ O Y N	1.2 Was the hazardous condition(s) recognized? If yes answer A and B. If no, proceed to 1.3.		Perform job safety analysis. Improve employee ability to recognize existing or potential hazardous conditions. Provide test equipment, as required, to detect hazard. Review any change or modification of equipment/tool(s)/ material.	
□ O Y N	A. Was the hazardous condition(s) reported?		Train employees in reporting procedures. Stress individual acceptance of responsibility.	
□ O Y N	B. Was the employee(s) informed of the hazardous condition(s) and the job procedures for dealing with it as an interim measure?		Review Job procedures for hazard avoidance. Review supervisory responsibility. Improve supervisor/ employee communications. Take action to remove or minimize hazard.	
□ O Y N	1.3 Was there an equipment inspection procedure(s) to detect the hazardous condition(s)?		Develop and adopt procedures (for example, an inspection system) to detect hazardous conditions. Conduct test.	
□ O Y N	1.4 Did the existing equipment inspection procedure(s) detect the hazardous condition(s)?		Review procedures. Change frequency or comprehensiveness. Provide test equipment as required. Improve employee ability to detect defects and hazardous conditions. Change job procedures as required.	
□ O Y N	1.5 Was the correct equipment/tool(s)/material used?		Specify correct equipment/tool(s)/ material in job procedure.	
□ O Y N	1.6 Was the correct equipment/ tool(s)/material readily available?		Provide correct equipment/tool(s)/ material. Review purchasing specifications and procedures. Anticipate future requirements.	
□ O Y N	1.7 Did employee(s) know where to obtain equipment/tool(s)/ material required for the job?		Review procedures for storage, access, delivery, or distribution. Review job procedures for obtaining equipment/tool(s)/material.	

Guide for Identifying Causal Factors & Corrective Actions

Figure 4-3. Guide for Identifying Causal Factors and Corrective Actions

	Causal Factors	Comment	Possible Corrective Actions	Recommended Corrective Actions
O □ Y N	1.8 Was substitute equipment/tool(s)/material used in place of the correct one?		Provide correct equipment/tool(s)/material. Warn against use of substitutes in job procedures and in job instruction.	
O □ Y N	1.9 Did the design of the equipment/ tool(s) create operator stress or encourage the operator error?		Review human factors engineering principles. Alter equipment/tool(s) to make it more compatible with human capability and limitations. Review purchasing procedures and specifications. Check out new equipment and job procedures involving new equipment before putting into service. Encourage employees to report potential hazardous conditions created by equipment design.	
O □ Y N	1.10 Did the general design or quality of the equipment/ tool(s) contribute to a hazardous condition?		Review criteria in codes, standards, specifications, and regulations. Establish new criteria as needed.	
O	1.11 List other causal factors in "Comment" column.			

PART 2 ENVIRONMENT

O □ Y N	2.0 WAS THE LOCATION OF EQUIPMENT/MATERIALS/EMPLOYEE(S) A CONTRIBUTING FACTOR? If yes, answer the following. If no, proceed to Part 3.

	Causal Factors	Comment	Possible Corrective Actions	Recommended Corrective Actions
O □ Y N	2.1 Did the location/position of equipment/material/employee(s) contribute to a hazardous condition?		Perform job safety analysis. Review job procedures. Change the location, position, or layout of the equipment. Change position of employee(s). Provide guardrails, barricades, barriers, warning lights, signs, or signals.	
□ O Y N	2.2 Was the hazardous condition recognized? If yes, answer A and B. If no, proceed to 2.3.		Perform job safety analysis. Improve employee ability to recognize existing or potential hazardous conditions. Provide test equipment, as required, to detect hazard. Review any change or modification of equipment/tools/materials.	
□ O Y N	A. Was the hazardous condition reported?		Train employees in reporting procedures. Stress individual acceptance of responsibility.	
□ O Y N	B. Was employee(s) informed of the job procedures for dealing with the hazardous condition as an interim action?		Review job procedures for hazard avoidance. Review supervisory responsibility. Improve employee/supervisor communications. Take action to remove or minimize hazard.	
□ O Y N	2.3 Was employee(s) supposed to be in the vicinity of the equipment/material?		Review job procedures and instruction. Provide guardrails, barricades, barriers, warning lights, signs or signals.	

		Causal Factors	Comment	Possible Corrective Actions	Recommended Corrective Actions
□ O Y N		2.4 Was the hazardous condition created by the location/position of equipment/material visible to employee(s)?		Change lighting or layout to increase visibility of equipment. Provide guardrails, barricades, barriers, warning lights, signs or signals, floor stripes, etc.	
□ O Y N		2.5 Was there sufficient work-space?		Review workspace requirements and modify as required.	
O □ Y N		2.6 Were environmental conditions a contributing factor (for example, illumination, noise levels, air contaminant, temperature extremes, ventilation, vibration, radiation)?		Monitor, or periodically check, environmental conditions as required. Check results against acceptable levels. Initiate action for those found unacceptable.	
O		2.7 List other causal factors in "Comment" column.			

PART 3 PEOPLE

O □ Y N	3.0 WAS THE JOB PROCEDURE(S) USED A CONTRIBUTING FACTOR? If yes, answer the following. If no, proceed to Part 3.6.

		Causal Factors	Comment	Possible Corrective Actions	Recommended Corrective Actions
□ O Y N		3.1 Was there a written or known procedure (rules) for this job? If yes, answer A, B, and C. If no, proceed to 3.2.		Perform job safety analysis and develop safe job procedures.	
□ O Y N		A. Did job procedures anticipate the factors that contributed to the accident?		Perform job safety analysis and change job procedures.	
□ O Y N		B. Did employee(s) know the job procedure?		Improve job instruction. Train employees in correct job procedures.	
O □ Y N		C. Did employee(s) deviate from the known job procedure?		Determine why. Encourage all employees to report problems with an established procedure to supervisor. Review job procedure and modify if necessary. Counsel or discipline employee. Provide closer supervision.	
□ O Y N		3.2 Was employee(s) mentally and physically capable of performing the job?		Review employee requirements for the job. Improve employee selection. Remove or transfer employees who are temporarily, either mentally or physically, incapable of performing the job.	
O □ Y N		3.3 Were any tasks in the job procedure too difficult to perform (for example, excessive concentration or physical demands)?		Change job design and procedures.	
O □ Y N		3.4 Is the job structured to encourage or require deviation from job procedures (for example, incentive, piecework, work pace)?		Change job design and procedures.	

		Causal Factors	Comment	Possible Corrective Actions	Recommended Corrective Actions
O		3.5 List other causal factors in "Comment" column.			
O ☐ Y N		3.6 WAS LACK OF PERSONAL PROTECTIVE EQUIPMENT OR EMERGENCY EQUIPMENT A CONTRIBUTING FACTOR IN THE INJURY? If yes, answer the following. If no, proceed to Part 4. Note: The following causal factors relate to the *injury*.			

		Causal Factors	Comment	Possible Corrective Actions	Recommended Corrective Actions
☐ O Y N		3.7 Was appropriate personal protective equipment (PPE) specified for the task or job? If yes, answer A, B, and C. If no proceed to 3.8.		Review methods to specify PPE requirements.	
☐ O Y N		A. Was appropriate PPE available?		Provide appropriate PPE. Review purchasing and distribution procedures.	
☐ O Y N		B. Did employee(s) know that wearing specified PPE was required?		Review job procedures. Improve job instruction.	
☐ O Y N		C. Did employee(s) know how to use and maintain the PPE?		Improve job instruction.	
☐ O Y N		3.8 Was the PPE used properly when the injury occurred?		Determine why and take appropriate action. Implement procedures to monitor and enforce use of PPE.	
☐ O Y N		3.9 Was the PPE adequate?		Review PPE requirements. Check standards, specifications, and certification of the PPE.	
☐ O Y N		3.10 Was emergency equipment specified for this job (for example, emergency showers, eyewash fountains)? If yes, answer the following. If no, proceed to Part 4.		Provide emergency equipment as required.	
☐ O Y N		A. Was emergency equipment readily available?		Install emergency equipment at appropriate locations.	
☐ O Y N		B. Was emergency equipment properly used?		Incorporate use of emergency equipment in job procedures.	
☐ O Y N		C. Did emergency equipment function properly?		Establish inspection/monitoring system for emergency equipment. Provide for immediate repair of defects.	
		3.11 List other causal factors in "Comment" column.			

PART 4 MANAGEMENT

O □ Y N	**4.0 WAS A MANAGEMENT SYSTEM DEFECT A CONTRIBUTING FACTOR?** If yes, answer the following. If no, STOP. Your causal factor identification exercise is complete.			

	Causal Factors	**Comment**	**Possible Corrective Actions**	**Recommended Corrective Actions**
O □ Y N	4.1 Was there a failure by supervision to detect, anticipate, or report a hazardous condition?		Improve supervisor capability in hazard recognition and reporting procedures.	
O □ Y N	4.2 Was there a failure by supervision to detect or correct deviations from the job procedure?		Review job safety analysis and job procedures. Increase supervisor monitoring. Correct deviations.	
□ O Y N	4.3 Was there a supervisor/ employee review of hazards and job procedures for tasks performed infrequently? (Not applicable to all accidents)		Establish a procedure that requires a review of hazards and job procedures (preventative actions) for tasks performed infrequently.	
□ O Y N	4.4 Was supervisor responsibility and accountability adequately defined and understood?		Define and communicate supervisor responsibility and accountability. Test for understandability and acceptance.	
□ O Y N	4.5 Was supervisor adequately trained to fulfill assigned responsibility in accident prevention?		Train supervisors in accident prevention fundamentals.	
O □ Y N	4.6 Was there a failure to initiate corrective action for a known hazardous condition that contributed to this accident?		Review management safety policy and level of risk acceptance. Establish priorities based on potential severity and probability of recurrence. Review procedure and responsibility to initiate and carry out corrective actions. Monitor progress.	
O	4.7 List other causal factors in "Comment" column.			

Developed by the National Safety Council
©2006 National Safety Council

SUMMARY OF CAUSAL FACTORS

General Category

Dates of Cases: From   To

SUBCATEGORIES · CATEGORY TOTAL   SUBCATEGORIES · CATEGORY TOTAL

1.0 WAS A HAZARDOUS CONDITION(S) A CONTRIBUTING FACTOR.

1.1 Defect(s) in equipment/tool(s)/material to hazardous condition(s).

1.2 Hazardous condition(s) not recognized.

  A. Hazardous condition(s) recognized, but not reported.

  B. Hazardous condition(s) reported but employee(s) not informed of known hazard and job procedures for hazard avoidance.

1.3 Lack of equipment inspection procedure(s) to detect hazardous condition(s).

1.4 Existing equipment inspection procedure(s) did not detect hazardous condition(s).

1.5 Correct equipment/tool(s)/material not used.

1.6 Proper equipment/tool(s)/material not readily available.

1.7 Employee(s) did not know where to obtain equipment/ tool(s)/material.

1.8 Substitute equipment/tool(s)/material used in place of proper one.

1.9 Equipment/tool(s) design created an operator stress or encouraged operator error.

1.10 General design or quality of equipment/tool(s) contributed to hazardous condition.

1.11 Other causal factors.

2.0 LOCATION/POSITION OF THE EQUIPMENT/MATERIAL/ EMPLOYEE(S) WAS A CONTRIBUTING FACTOR.

2.1 Location/position of equipment/material/employee(s) contributed to hazardous condition(s).

2.2 Hazardous condition(s) not recognized

  A. Hazardous condition(s) recognized, but not reported.

B. Hazardous condition(s) reported, but employee(s) not informed of job procedure for dealing with the hazardous condition as an interim action.

2.3 Employee(s) should not have been in the vicinity of the equipment/material.

2.4 Hazardous condition(s) not seen by employee(s).

2.5 Workspace insufficient.

2.6 Environmental condition(s) a contributing factor.

2.7 Other causal factors.

3.0 JOB PROCEDURE WAS A CONTRIBUTING FACTOR.

3.1 No written or known job procedures.

  A. Job procedures existed, but employee(s) did not anticipate factors contributing to the accident.

  B. Job procedures existed, but employee(s) did not know them.

  C. Employee(s) knew job procedures, but deviated from them.

3.2 Employee(s) not capable of performing job.

3.3 Task in the job procedures too difficult to perform.

3.4 Job structured to encourage deviation from job procedures.

3.5 Other causal factors.

3.6 LACK OF PERSONAL PROTECTIVE EQUIPMENT OR EMERGENCY EQUIPMENT WAS A CONTRIBUTING FACTOR IN THE INJURY.

3.7 Appropriate personal protective equipment (PPE) not specified.

  A. PPE specified, but not available.

  B. PPE specified, but employee(s) did not know that PPE was required.

Figure 4-4. Guide for Identifying Causal Factors and Corrective Actions

	SUBCATEGORIES	CATEGORY TOTAL	

C. PPE specified, but employee(s) did not know how to use or maintain PPE.

3.8 PPE not used properly.

3.9 PPE inadequate.

3.10 Emergency equipment not specified for this job.

A. Emergency equipment specified, but not readily available?

B. Emergency equipment specified, but was not used properly.

C. Emergency equipment specified, but did not function properly.

3.11 Other causal factors.

4.0 A MANAGEMENT SYSTEM DEFECT WAS A CONTRIBUTING FACTOR.

4.1 Failure by supervision to detect, anticipate, or report a hazardous condition.

4.2 Failure by supervision to detect or correct deviation from job procedure.

4.3 Failure to conduct a supervisor/employee review of hazards and job procedures for tasks performed on infrequent basis.

4.4 Supervisor responsibility and accountability not defined or not understood.

4.5 Supervisor not adequately trained to fulfill assigned responsibility in accident prevention.

4.6 Failure to take a corrective action on a known hazardous condition(s).

4.7 Other causal factors.

Developed by the National Safety Council
©2006 National Safety Council

EMERGENCY PLANNING

CHAPTER 5

Emergencies in aviation ground operations include aircraft incidents, fuel spills, fires, hurricanes, tornadoes, floods, earthquakes, hijackings, bombs, bomb threats, dangerous goods and airport civil defense. Many of these emergencies require immediate response by specialized teams, such as the fire/rescue brigade or bomb and security team. The Department of Homeland Security in the United States has taken an active role in responding to these events.

Regardless of the size or type of organization, management is responsible for developing, operating and maintaining a flexible emergency action program that will provide:

- Maximum protection for passengers, employees, and the general public
- Maximum protection of equipment, facilities and property
- Immediate and effective response to incidents or emergencies
- The ability to take prompt action in saving lives, and preventing suffering or minimizing destruction or damage to property

This chapter provides basic information to help develop emergency action plans. Each area in an organization must prepare detailed operating procedures, which are to be part of the overall emergency action plan. The operating procedures could be in checklist form for ease in both instruction and use. These procedures must:

- Instruct how to notify key personnel in an emergency.
- Assign responsibilities and/or tasks by position title.
- Define operating locations.

- Outline procedures/instructions on how each task will be accomplished. Instructions must be detailed enough to enable the individuals or unit to function properly.

All emergency programs should incorporate all or most of the following: a printed plan, company-wide planning, command headquarters, protection of records, personnel shelter areas, communications systems, fire prevention and protection, property protection and security, first aid and medical, rescue and salvage, warden service and evacuation, transportation, mutual aid plans, local services, and program testing procedures. Training of personnel is extremely important.

Preparedness is insurance against sudden disaster. While it may not always be possible to prevent a disaster from happening, it is possible to minimize its effects by knowing just what to do and then doing it promptly.

Policy and procedures

Regardless of the type of incident, the first priority is the rescue of all involved; the second is the protection of aircraft, buildings and equipment.

It is strongly recommended that each airport, air carrier, and corporate or fixed base operator have its own emergency plan(s) so the procedures can be put into effect quickly. These plans should be coordinated with airport management to make sure no conflicts exist. Although incidents and emergencies vary in nature, responses will be essentially the same.

Response to emergencies can be divided into phases to readily identify the actions required by each person in each phase. The phases are:

- *Notification.* Any agency or person receiving or having information of an actual or potential incident/incident or disaster will immediately relay this information to the predesignated communication center for appropriate action.
- *Response.* Each response agency must be prepared to act individually and should maintain contact with

the communication center.

- *Recovery*. Recovery is a complex and time-consuming operation. Experience has shown that from one week to several months may be required to make a complete recovery following an incident or disaster. A recovery plan must be developed based on the requirements/findings of all response agencies.

Incident plan. Items to be considered in an incident plan are:

- *A statement of management policy*. Incident response procedures should outline in broad detail the policy, action, organization, and responsibilities of individuals and groups who should govern the handling of all major incidents.
- *Periodic review*. Responsible individuals and groups should meet to review and update the incident procedures, as well as to discuss possible problems.
- *Coordinators*. These people would handle the following:

  - Communications
  - Medical assistance
  - Corporate liaison officers
  - Relative centers
  - Identification bases
  - Scene
  - Engineering and maintenance
  - Public relations/news media
  - Security
  - Safety

Establishment of notification/communication responsibilities. A detailed emergency plan would include provisions for notifying airline personnel, as well as outside agencies. Items in the plan might include:

- *Establishment of a central communication facility*. The overall assigned coordinator can keep in constant communication with all other coordinators as listed above.
- *Establishment of a notification list*. Principals assigned incident responsibilities are to be notified

with a minimum of delay. The notification listing should contain current names and telephone numbers (office and home) of individuals assigned specific tasks. Copies of the listing should be readily available in the communications center and included in the on-scene coordinators response kit.

- *Establishment of a method for notifying:*

  - Civil Aeronautics Board (CAB), Federal Aviation Administration (FAA), Federal Bureau of Investigation (FBI), Royal Canadian Mounted Police (RCMP), Transport Canada (TC), International Police (Interpol), Scotland Yard, or other agencies having jurisdiction
  - Medical assistance
  - Postal authority
  - Package forwarding agencies
  - Courier services (if agents are involved)
  - Department of Defense (if security cargo is involved)
  - Appropriate military services (if service personnel are involved)

Establishment of control responsibilities

- *Command/direction*. The airfield manager or fire chief (someone in complete charge) will direct and control all airfield operations relative to this plan. Persons having knowledge of an impending or actual disaster/major incident will immediately notify the appropriate designated individual (or position title).
- *Control* by sirens, loudspeaker, telephone, intercom, public address system, radio or runners.
- *Establish a disaster response group* headed by the team chief/on-scene director described above. This group should consist of selected representatives from all airfield support activities who are qualified to take action in their respective functional areas and provide technical support to the team chief on particular facets of the situation. The disaster response group will include, but not be limited to, the following members:

  - On-scene director
  - Fire chief

- Security
- Medical
- Maintenance

Establishment of specific procedures

Establish procedures and designate individuals to secure and distribute the following information:

- Trip number (if air carrier)
- Date and time of incident
- Aircraft type (and plane number, both company and "N" numbers, if air carrier)
- Location of incident
- Crew names and domiciles
- Passenger lists
- Cargos aboard such as mail, express, airfreight, baggage, company materials and hazardous materials. Advise on-scene coordinator and identification coordinators immediately of any hazardous cargoes (radioactive, for example) or extremely valuable shipments.

Establish procedures and designate individuals to arrange for:

- Protection of aircraft, cargo, passengers' belongings and company equipment
- News media control
- Crowd control
- Identification of authorized personnel
- Roadblocks as required
- Services of ambulances and hospitals
- Accommodations for personnel secured from other areas, relatives, etc.
- Transportation of personnel groups and equipment to the scene
- Food and special clothing
- Special equipment, such as bulldozers and cranes, for crash removal (coordination is especially important where Army units are located in the vicinity)
- Special personnel needs such as drivers, mountain climbers, water rescue personnel, etc. (Where Army and Air National Guard units are located in the vicinity, coordination should be established for emergency use of security guards to prevent looting and provide

safekeeping of wreckage; coordination should also be established for possible usage of unit paramedics)
- Additional firefighting and rescue equipment
- Assistance for the injured
- Assembly point for the uninjured
- Center for relatives
- Identification of passengers and belongings
- Disposition of bodies
- Temporary facilities to serve as hospitals and morgues
- Communication equipment (portable radios, transceivers and base station)
- Emergency lighting

Availability of nearby facilities. The following contacts should be readily available. The type of assistance they can render should be determined in advance.

- Hospital
- Ambulances
- Firefighting and rescue vehicles
- State police and other enforcement agencies
- Maps and sketches for distribution to authorized individuals to use as guides to an incident scene
- Car and truck rentals
- Special equipment rentals
- Civil defense
- Government agencies

Emergency salvage kit. Consideration also should be given to establishing an emergency salvage kit if possible, or at least know the local source where the following items will be readily available:

- Supply of stationery items (paper, pencils, envelopes, carbon paper, folders, etc.)
- Rope, at least 300 m (1,000 ft), to be used in securing area, hoisting equipment, tying down tarps and other tasks
- Miscellaneous: gloves (both rubberized and leather palm), hatchets, hammers, nails, hunting knives, blankets, flour or sugar sacks (for small items and personal effects), flashlights and spare batteries, tags with wire attached, block and tackle body bags, stretchers, axes, vacuum jugs, water cans, rakes,

shovels and picks, coveralls, cameras and film, small hand tools, containers for fuel and oil samples, marking pencils, plastic bags, stakes for securing area, and adequate footwear for area (boots, overshoes, etc.)

- Other items (such as cranes and barges) may be desirable and essential; the need is based on local terrain and conditions.

FUEL SPILLS AND FIRES

Spills of either gasoline or kerosene present a potentially serious fire hazard to personnel, aircraft, and equipment each time they occur (for a description of a hazardous material incident see Chapter 3, "Hazardous Materials and Harmful Physical Agents"). Therefore, every effort should be made to prevent them.

Controlling spills

The following are recommendations to minimize the number of incidental spills:

- Develop a spill control procedure and establish a reporting protocol.
- Fuel servicing personnel should be thoroughly trained. They should observe proper grounding and bonding procedures. During overwing fueling, the fueler must never leave the hose nozzle unattended. The nozzle trigger is never to be wedged or tied in an open position. The fueler also should devote full attention to the job, making frequent checks of fuel level to prevent overfilling. During underwing fueling, the fueler also must understand the operation of the automatic shutoff in the aircraft fueling system and be able to monitor its operation at all times.
- All fueling equipment is to be properly maintained. It should be inspected daily and repaired as required.
- Every spill should be reported and investigated so that remedial action can be taken.
- If a fuel spill occurs, it should be treated as a potential fire source and the following action taken (note that if these are in conflict with local airport regulations, the local regulations take precedence):

1. Stop fuel flow by closing the emergency shut-off valve immediately. Depending upon the nature of the spill, it may be necessary to evacuate the aircraft.
2. Notify the airport fire department (except for small spills as defined below). Supervisory personnel should be notified to determine whether operations in progress may either be continued safely or should be halted until corrective measures have been taken.

- All ramp staff should know where emergency fuel flow shutoff valves are located and how to shut off valves.

Small spills of less than 0.5 m (20 in.) in any direction normally are considered minor. Other than normal standby of ramp personnel with fire extinguishers, no other precautions are considered necessary.

Volatility of fuel is a major factor in spill hazard severity. Aviation gasoline and other low flashpoint fuels at normal temperatures and pressures will give off vapors capable of forming ignitable mixtures with the air near the surface of the liquid. This condition does not normally exist with kerosene fuels. Each spill will have to be treated individually because of such variables as wind and weather conditions, equipment arrangement, personnel, and available emergency equipment.

It may be necessary to evacuate the aircraft if it appears the spill may present a serious fire exposure to the aircraft. Personnel should not walk through the spill area.

Motorized equipment within spill areas should not be started before a spill is cleaned up. However, if the engine is already running the equipment may be moved if safe to do so.

Motorized equipment outside a spill area should be moved away or shut down as quickly as possible. If the fuel servicing equipment is moved, it should be determined that the fuel hose or pipe to the aircraft has been disconnected and stowed.

Large fuel spills should be blanketed with foam. The spill should then be cleaned up and residues handled as hazardous waste in accordance with local regulations. Fuel should not be washed down sewers or drains. If no alternative exists, the use of sewer or drains for elimination of fuel spills should be approved by authorities having jurisdiction and performed under their supervision. If this is done, the sewers should be flushed with large quantities of water.

Following a spill, depressed areas such as flap wells or wing sections of the aircraft involved should be inspected for accumulations of fuel or fuel vapors. Any fuel discovered should be removed.

All cargo should be inspected after a spill. If any cargo is contaminated with fuel, it should not be placed aboard until the fuel is removed or evaporated.

The aircraft should be removed from the fuel area before its engines are started, unless the spill has been completely eliminated.

Controlling fires

If a fire does occur after a fuel spill, the following procedures are recommended and should be performed simultaneously (if sufficient personnel are available):

- The airport fire department and appropriate company management personnel should be called immediately.
- If an aircraft, fuel truck, or spilled fuel catch fire, the aircraft should be immediately evacuated. If fire prevents the use of the normal passenger loading stands, passengers should evacuate through the emergency exits.
- All equipment should be available to fight the fire. All members of a fueling crew in particular, and all ramp personnel in general, should be thoroughly trained in the use of fire extinguishers. Their training should include practice drills in extinguishing flammable liquid fires.
- If possible, the fuel truck should be moved from the fire area, following the precautions discussed above.

BOMB THREATS

Bomb threats usually are received in one of two ways:

- Casual remarks by passengers on aircraft or at terminals
- Anonymous phone calls or letters

Of the two, the latter is usually considered more serious. Both require investigation, however. The FBI or other cooperating agencies should be informed of all bomb threats.

Procedures

The following procedures are usually involved in bomb threat cases:

- Any person or agency receiving or having information regarding a bomb threat should record in detail all information concerning the threat and immediately pass it to the appropriate designated agency by the most expeditious means.
- The person receiving the call directly from the individual making the threat should solicit as much information as possible from the caller. If possible, the person receiving the call should leave the line open and use another line to request the telephone operator to initiate tracer action. The person receiving the call should attempt to determine the following:
- Any accents or speech peculiarities of the caller
- Whether the caller is male or female
- If the call is local or long distance
- Any additional information that may assist authorities in their investigation
- Record the time the call was received
- Flights may be requested to return to the terminal. The aircraft may be removed to an isolated area. Passengers' baggage may then be removed and searched, usually by local authorities and the operating airline. Authorities or airline personnel will advise passengers on their course of action. The FBI normally interrogates all passengers.
- The aircraft may be thoroughly searched, usually by personnel of the operating airline.
- Certain military installations have dogs specially

trained to sniff out most types of explosives. If a military installation is near, this possibility should be explored.

- Clearance for flight should be given only after all involved authorities have been satisfied that no bomb is on the aircraft.

NATURAL DISASTERS

Airline emergency procedures should include action plans for natural disasters such as hurricanes, tornadoes, floods, and earthquakes.

Weather bureaus are now able to detect and track hurricanes so that ample warning generally can be given for protection of property and for evacuation from threatened areas. Many airline weather units also can provide valuable warning information.

Establish severe weather warning procedures and checklists. Lead times for severe weather forecasts should be established and identified in order to complete required precautionary tasks. Lead times vary by locale, so individual airports should establish times or distances for the following weather phenomena:

- Tornadoes/hurricanes/typhoons sighted within x kilometers or nautical miles of the airport
- Hail
- Surface winds 35 knots (18 m/s) or greater
- Severe electrical storms
- Freezing precipitation
- Heavy snow accumulation
- Thunderstorms

With tornadoes, however, it sometimes is not possible to provide as much advance warning or to anticipate the specific strike area as it is with other windstorms. Therefore, companies should be prepared with action plans designed to protect their employees, passengers and property.

In North America, many areas with seismic activities are found along the Pacific Coast. Earthquakes occur without warning and generally affect an entire community or a large area. Therefore, assistance from community services may not be available and companies and/or airports may be solely dependent upon their own action plans.

Action planning

Emergency action plans for all contingencies should consist of three parts:

1. A thorough preparedness program developed in anticipation of an emergency condition
2. A specific plan of action including definite assignments of responsibility during emergency conditions
3. A "clean-up" or salvage procedure to follow emergency conditions so that normal operations may be resumed as quickly as possible.

Copies of each response agency's action plan must be consolidated into one master document that is readily available to the command and control center.

Develop a procedure for evacuating personnel to a safe place. In the event of an earthquake, evacuation should take place after the earthquake has subsided. During the earthquake, immediate protection under the nearest available cover should be sought. Personnel should avoid standing near windows or collapsible structures.

Some specific items to consider in developing the procedure are:

- Shelter areas
- Emergency lighting
- Alarm systems
- Protection of aircraft; consider dispersal procedures and tie-down facilities
- Emergency water supply for firefighting services
- Protection of vital records (and duplication, if necessary)
- Design and construction of buildings in accordance with all applicable codes so that they can better withstand destructive forces of nature
- Procedures for shutting down and restoration of utilities such as steam, electricity, water and gas
- Assignment and training of specific crews for firefighting, rescue service, guard service, first aid and medical service, demolition and repair, transporta-

tion, investigation, public relations, communications, and warden services.

Setting up an action committee

An action committee should also be appointed. This group is responsible for developing and executing all emergency activities that have been specified by the emergency committee.

A sufficient number of members should be appointed to the on-scene disaster control group to ensure full coverage on a 24-hour operational basis. A typical on-scene disaster control group should consist of the following membership:

- On-scene director/chairperson
- Fire/rescue
- Medical
- Security
- Public relations/news
- Maintenance/engineering
- Safety
- Photography
- Legal
- Flight operations

The action committee is an "on-the-scene" committee with the following responsibilities:

- **Police services.** These include espionage and sabotage control, prevention of panic and mass hysteria, traffic control to shelter areas plus evacuation of plant, policing after an attack to prevent looting and to protect company property generally, and plant security services.
- **Records control.** Important records at all plants should be protected in storage locations – such as vaults. Documents that are indispensable to company operations should be duplicated and stored in remote locations.
- **Health services.** First aid training should be given to select company personnel. Facilities for transportation of the injured should be provided – both within the plant and from the plant to the hospital. First aid supplies should be provided, as well as food and water supplies.

- **Personnel rescue, facilities rehabilitation and salvage operations.** These functions most logically would be handled by the group normally responsible for routine maintenance of the facilities and equipment. Duties of this group would include releasing trapped and injured persons and transferring them to first aid; setting up a motor pool of all available equipment and providing drivers; shutting off broken gas lines, water mains, and power lines and restoring services as soon as possible; cleaning up buildings and grounds as soon as possible; working with the radiation monitoring crew in removing radioactive dust where necessary; and inspecting buildings and facilities for damage.
- **Personnel protection.** The safest areas (from blast and fallout) in the structures should be designated shelter areas. Depending upon the nature of the emergency, personnel should either be directed to the shelters or ordered to evacuate the premises in accordance with prearranged plans. Wardens should be assigned to assist personnel.
- **Fire prevention and extinguishment.** Brigades trained in firefighting and fire control should be established. After receiving an alarm, they should respond and attempt to extinguish fires using plant fire appliances. Other associated activities – closing fire doors, securing flammable liquids, and salvage operations – should be included in the activities of this group.
- **Radiation monitoring.** Proper instruments should be obtained and a crew trained to measure radioactivity following an emergency. Controls should be set up to protect employees from entering radioactive areas. (This crew also would work with the decontamination crew to clean up radioactive areas.)
- **Communications.** A means should be provided to receive alert messages signalling an emergency and for transmitting this alert to plant personnel so appropriate action can be taken.

References

National Safety Council:
Accident Prevention Manual for Business and Industry, Administration and Programs, 11th ed. 1997.
On-Site Emergency Response Planning Guide. 1999.

FIRE PREVENTION AND
PROTECTION

CHAPTER 6

Fire prevention and protection starts in the planning of facilities when fire-resistant construction and fire suppression systems (e.g., automatic sprinklers and chemical fire systems) are incorporated into the design. Facilities should be planned using appropriate national fire standards, consulting with insurance engineers and underwriters, and with the local authority having jurisdiction. Fire protection requirements should be met or exceeded where the codes, which are minimum standards, do not provide for protection adequate to the risk.

This chapter presents basic information on the nature of fires and on incipient stage fire protection programs. It also gives some basic standards for fire protection in several aviation-specific areas such as maintenance operations.

NOTE: FOR MORE DETAILED STANDARDS, CHECK THE U.S. NATIONAL FIRE CODES OR YOUR OWN COUNTRY'S FIRE STANDARDS FOR SPECIFIC APPLICATIONS AND REQUIREMENTS.

THE BASIS OF THIS CHAPTER DEALS WITH FIRST AID FIREFIGHTING. PROPERLY EDUCATED AND TRAINED PERSONNEL WITH ADEQUATE EQUIPMENT CAN CONTROL OR EXTINGUISH SMALL FIRES PRIOR TO THE ARRIVAL OF PROFESSIONAL FIREFIGHTERS. THE PRIMARY GOAL IS LOSS PREVENTION.

FIRE PROTECTION – GENERAL INFORMATION

The National Fire Protection Association's (NFPA) National Fire Codes and other national and local standards have been developed over the years, in many cases as the result of actual fire experience. According to NFPA, "The intent of these codes is to prescribe minimum requirements necessary to establish a reasonable level of fire safety and property protection from fire and explosion." Some codes apply specifically to aviation facilities and operations, while others are more general (see References)

COMPONENTS OF FIRE

Fire requires four elements to exist. Fire must have fuel, oxygen, heat, and a chemical reaction to exist (Figure 6-1). Without any one of these elements fire will not occur. Remove any one of these elements and the fire will extinguish.

Basic elements

Removing the heat of a fire is, for some types of combustibles, one way of putting out the fire. A good example of this is the use of water as a solid stream, fog or steam to put out a Class A fire. (Class A: flammable solids such as wood, paper products, rubber, textile and plastics; fires that leave ash.)

Removing the oxygen from a fire can be done in several ways. One can blanket the fire with a heavy,

Figure 6-1. Oxygen, heat, fuel, and a chain reaction are necessary for a fire.

noncombustible gas such as CO_2, or cover the fuel with a blanket of foam, which also cools the fire. Where metal fires or Class D fires occur, the burning metal is often blanketed with a mixture of salt and graphite that forms a hard crust and excludes oxygen.

Interrupting the chemical reaction of a fire is a very common method used in fighting fire with dry chemical agents. The dry chemical agent chemically interrupts the combustion reaction to stop fires. In the past halons were used for the same reason, but these agents are members of the CFC family and are being phased out.

Removing the fuel, if possible, will extinguish a fire. In flammable liquid and gas fires sometimes a simple closing of a valve shuts off the source of the fuel from the burning spill and will allow the fire to burn out.

Classes

Even though there are five to seven classes of fire, the most common types encountered in aviation are:

Class A Flammable solids such as wood, paper products, rubber, textiles, and plastics (fires that leave ash)

Class B Flammable and combustible liquids and gases

Class C Electrical (usually a Class A fire with an energized electrical circuit)

Class D Metals such as magnesium (aircraft wheels

FIRE PROTECTION

The ramp areas present a wide range of hazards to be considered in evaluating fire equipment needs. In addition to the normal aircraft related potential, with flammable liquid, and electrical hazards, the aircraft servicing equipment also presents its own unique exposures. Different aircraft types present unique firefighting challenges. If aircraft types are mixed on the ramp, the extinguisher selected shall be suitably sized for the largest aircraft. Refer to NFPA or applicable local regulations for specific recommendations.

Vehicles present a potential hazard to themselves and to the aircraft they are servicing. Vehicles that are not easily moved away from the aircraft in the event of a fire should be equipped with fire extinguishers. The recommendation is a minimum of one 20 lb. (8.2 kg) extinguisher, UL rating of 80 BC, using Purple K dry chemical, mounted on each vehicle.

Some ground support equipment have built-in internal fire suppression system. Know your equipment emergency fuel shutoff and fire system operation. If the combustible component is not accessible a port hold should be provided to allow a means of applying firefighting agents to the internal areas of the equipment. This access port should be clearly identified.

Special precautions for ramp area

Fuel handling, outside storage of materials, and rubbish pose special fire hazards in ramp areas and require special fire protection measures. See your own country's standards or the U.S. National Fire Codes on all three kinds of operations.

Fuel handling. Gasoline and gasoline-blended fuels can ignite at extremely low temperatures if an ignition source is present. The temperature range in which it is possible to have flammable vapor-air mixtures is quite wide and covers the normal ambient temperature range. For example, at sea level in a gasoline storage tank, the temperature range could be from -46° C to 74° C (-50° F to 165° F). The temperature range for any one fuel, however, would be much narrower.

Autoignition temperature is the minimum temperature of a substance required to initiate or cause self-sustained combustion, independent of any sparks or other means of ignition. Under some conditions, the autoignition temperature of jet fuels has been measured at 250° C (475° F). This temperature may exist for a considerable period in turbine engines after shutdown or on brake surfaces following hard use. (These figures were derived from laboratory tests; in field conditions the ignition temperatures may be higher.)

Static discharge is one source of ignition, and jet fuels are more prone to acquire a static charge than aviation gasoline.

Fueling and defueling of aircraft should be accomplished in accordance with procedures outlined in NFPA codes or equivalent national standards. The procedures are designed to reduce static electricity and other sources of ignition and limit the opportunity for fuel fires. (See References.)

Storage areas

Outdoor storage of materials that are susceptible to ignition by sparks or flying brands requires special measures. The following points should be considered to reduce the hazards of materials stored outdoors:

- Adequate fire and security protection of the storage areas should be in place.
- Well-maintained, all-weather roads should be available for fire apparatus.
- There should be sufficient clear space between any buildings and the place where combustibles are stored, as well as between any roads and other storage piles.
- Clear aisles should be maintained between storage piles, as well as between the piles and buildings and between the piles and the boundary line of a storage site. Driveways in the storage areas should be of sufficient width to permit fire equipment to travel to all parts of the storage area.
- The storage area should be surrounded by a security fence to reduce access by unauthorized personnel. There should be an adequate number of gates in the fence to permit ready access by fire equipment, and the area shall contain signs or placards listing hazardous contents.
- Any buildings in or near a storage area should be of noncombustible construction with adequate clearance to the nearest storage pile. Buildings that house hazardous operations such as spray painting and fuel control testing should be more than 50 ft (15m) from the nearest pile. Explosion vents, blower outlets, or similar devices should not be directed toward the storage area.
- Storage sites should be kept free of unnecessary combustible materials, to include weeds, trees, or grass. Regular cleanup is essential.
- There should be adequate lighting, in compliance with the National Electrical Code. (See References.)
- Public water systems should be adequate, with hydrants suitably located.
- Braziers, salamanders, portable heaters and other open fires should not be used in a storage yard.
- Tarpaulins, used for the protection of stored material, should be made of approved flame-resistant material.
- Firefighting equipment should be installed to meet the current requirements of the authority having jurisdiction.

Handling rubbish. Rubbish should be collected around an airport. Waste materials should be sorted into appropriate approved containers. This rubbish should be disposed of daily in a manner that ensures elimination of possible fire hazards.

Types of rubbish should be segregated according to the degree of compatibility and method of disposal. Rubbish should be handled in an orderly manner and not be allowed to accumulate where it may expose personnel, buildings or piled stock to fire hazards.

Planned burning of rubbish outdoors is one method of disposal but – apart from the smoke nuisance – may violate local or national clean air standards.

FIRE INSPECTIONS

A formal fire inspection program should be instituted and should include the periodic checks of buildings, building utilities, shop and hangar equipment, storage facilities, fuel storage areas, vehicles and the fire equipment provided to protect these areas. In order for the program to be effective, there should be unannounced, random inspections, as well as periodic ones.

A thorough fire inspection program should go beyond mere checks of facilities: It should include plans for eliminating or reducing fire hazards. An individual trained in fire inspection procedures should review potential hazards that occur in normal operations. Those engaged in routine operations, who may have production and schedules uppermost in their minds, may tend to accept hazards as an inherent part of the task. The person inspecting may find that some hazards can be eliminated or controlled. Some controls to consider are:

- Ways to improve housekeeping practices
- Better methods for handling flammable liquids
- Methods to reduce combustible material in storage or in process
- Improvements in emergency handling procedures

A designated person should conduct a fire prevention self-inspection at least monthly. Some buildings, operators, and processes on an airport may require daily inspections, while others may require weekly inspections. Use of an inspection checklist often is helpful (Figure 6-2). Using the sample list for each item the person conducting the self-inspection merely checks Yes," "No" or "Not applicable." The completed self-inspection checklist normally is forwarded to the airport fire marshall or a similar authority.

The airport local fire department, insurance underwriters, and government authorities having jurisdiction over the airport also undertake periodic inspections. Critical items, such as boilers and heating systems, electrical systems, pressure vessels, and flammable liquid or gas storage, should be given careful attention to ensure compliance with applicable codes. It is desirable for an atmosphere of cooperation to develop between inspecting authorities and operators.

All firefighting equipment should be inspected, tested, and maintained according to NFPA codes; and inspection records and their results maintained.

FIRE TRAINING

The action taken in the first few minutes of a fire or other emergency is of critical importance. Therefore, all personnel should be instructed in the procedures to follow prior to the arrival of professional firefighters. The instruction should include: fire alarms, evacuations, and incipient stage fire equipment operation, usage and capabilities.

Personnel should be considered first aid or incipient stage firefighters only, and should know the capabilities and limitations of their equipment.

NOTE: IF AN EMPLOYER CHOOSES TO FORM A FIRE BRIGADE MADE UP OF EMPLOYEES, THE STRICT REQUIREMENTS OF NFPA AND THE NATIONAL AUTHORITY HAVING JURISDICTION ON THIS SUBJECT SHALL BE FOLLOWED.

Instruction on calling for assistance should include how to use a fire alarm, where alarms ring to, and how and to whom to communicate additional information about an emergency. The activation mechanism of the fire alarm should be described and demonstrated. The importance of immediately calling for fire department assistance should be stressed. A responsible person from each area should be designated to meet and give directions to the fire department.

NOTE: THE FIRE STATION OR RESPONSE TEAM WITH FIRST RESPONSE RESPONSIBILITY FOR YOUR FACILITY SHOULD BE GIVEN REGULAR TOURS OF THE AREA (FIGURE 6-3). THEY SHOULD BE FAMILIAR WITH THE HAZARDS, ACCESS POINTS, BUILDING CONSTRUCTION, UTILITIES AND FUEL SHUTOFF POINTS, AND STANDPIPE AND FIREFIGHTING EQUIPMENT INCLUDING FIRE SYSTEMS PROTECTING THE BUILDING.

Instruction on extinguishing fires should include the location and use of portable fire extinguishers, and the correct type to use for each of the classes of fire. The instruction should include instruction hands on use of extinguishers in training fire situations, if possible.

Evacuation routes and assembly areas should be carefully mapped out and rehearsed. Alternate routes in case of smoke or obstacles should be planned and explained. Employees should be instructed never to enter a smoke-filled area. Assembly areas should be designated so that employees can be accounted for after evacuation.

FIRE INSPECTION CHECKLIST

Emergency Preparation
- ☐ Fire organization posted
- ☐ Fire drill held regularly
- ☐ Fire exits well marked and unobstructed
- ☐ Fire alarms well marked and unobstructed
- ☐ Aisles and stairs clear
- ☐ Evacuation procedures posted
- ☐ Procedures for handling fuel spills posted
- ☐ Emergency equipment well marked, in place and ready for use

Plant and Buildings
- ☐ Hydrants and water supply checked and serviceable
- ☐ Sprinkler system, checked and serviceable
- ☐ Foam and CO_2 systems, checked and serviceable
- ☐ Fire doors, checked and serviceable
- ☐ Fire extinguishers well marked, in place, checked and tagged
- ☐ Hose stations, checked and serviceable
- ☐ Electrical circuits identified, enclosed and provided with proper overload protection
- ☐ Gas systems, checked and serviceable
- ☐ Fuel pumping equipment, in good condition and free of leaks (extinguishing equipment adequate and available)

Production Equipment
- ☐ Spray booths, clean and properly ventilated and sprinkler heads protected from overspray
- ☐ Power tools and accessories, wiring in good condition
- ☐ Pressurized bottles, properly connected and secured
- ☐ Mobile equipment extinguishers well marked, in place and serviceable
- ☐ Powered equipment, properly grounded
- ☐ Test equipment, free of leaks, wiring in good condition.

Management
- ☐ General housekeeping, adequate
- ☐ Changes (alterations, processes, methods and procedures) cleared with Fire Marshal
- ☐ Floors, clean and free of flammable fluid spills
- ☐ Storage of material, orderly and in accordance with regulations*
- ☐ Aircraft fueling, in accordance with regulation*
- ☐ Spray painting, in accordance with regulations*
- ☐ Welding and other open flame operations, in accordance with regulations*
- ☐ Ramp and grounds, clean and free of debris
- ☐ Proper disposal of soiled shop towels and rags

* Separate checklists should be made for these operations.

Figure 6-2. To facilitate the process of fire inspection, the inspector can use a checklist like the one above.

Fire drills

Fire drills are a means to rehearse what has been taught and learned during fire training. Drills provide the assessment of the effectiveness of the training, and point out weaknesses in the plan and system needing corrective action. Fire drills should be carefully planned and monitored.

- Drills should be planned so that an actual fire condition can be simulated.
- Observers should be posted at strategic points throughout the area to review and time the actions of personnel and firefighting teams (if used).
- The drill may be conducted without warning to reproduce the conditions and reactions that are likely to exist in a real emergency.
- Regular drills should serve to make employees' reactions to the alarm automatic and to make it possible to cover different emergency conditions in a reasonably short cycle.
- Weaknesses in existing procedures should be recorded carefully so that improvements can be made.
- For each drill, a potential hazard should be selected and how it affects emergency procedures should be analyzed. For example, observers can be posted in main corridors to warn people being evacuated that "this corridor is theoretically on fire and an alternate means of evacuation should be found."

Figure 6-3. Shown are firefighters wearing heat protective fire proximity suits.

Photo courtesy of Memphis AARF

References

Factory Mutual System. *Aircraft Fueling Up-to-Date*
Industrial Risk Insurers. *Recommended Good Practice for Protection of High Piled Stock.*
National Fire Protection Association:
  Fire Brigade Training Manual.
  NFPA Inspection Manual.
  Standards:
     Aircraft Fueling, NFPA 407, 1990
     Aircraft Hand Portable Fire Extinguishers, NFPA 410, 1999
     Aircraft Maintenance, NFPA 410, 1999
     Aircraft Rescue and Fire Fighting Operations, NFPA 402, 1996
     Aircraft Rescue and Fire Fighting Services at airports, NFPA 403, 1998
     Airport Terminal Buildings, Fueling Ramp Drainage, and Loading Walkways, NFPA 415, 1997
     Blower and Exhaust Systems for Dust, Stock, and Vapor Removal or Conveying, NFPA 91, 1992
     Construction and Protection of Aircraft Loading Walkways, NFPA 471, 1992
     Construction and Protection of Airport Terminal Buildings, NFPA 416, 1987
     Evaluating Foam Fire Fighting Equipment on Aircraft Rescue and Fire Fighting Vehicles, NFPA 412, 2003
     Exhaust Systems for Air Conveying of Vapors, Gases, Mists and NonCombustible Particulate Solids, NFPA 91, 1999
     Fire Protection for Archives and Record Centers, NFPA 232A, 1995
     Installation of Air Conditioning and Ventilating Systems, NFPA 90A, 1999
     Life Safe Code, NFPA 101, 2000
     Low Expansion Foam, NFPA 11, 1998
     Master Planning Airport Water Supply Systems for Fire Protection NFPA 419, 1992
     Motor Freight Terminals, NFPA 513, 1998
     National Electrical Code, NFPA 232, 1995
     Portable Fire Extinguishers, NFPA 10, 1998
     Powered Industrial Trucks, Including Type Designations, Areas of Use, Conversion, Maintenance, and Operations, NFPA 505, 1999
     Protection of Records, NFPA 232, 1991
     Rack Storage of Materials, NFPA 231C, 1991
     Repair Garages NFPA 88B, 1997
     Removal of Smoke and Grease-Laden Vapors from

Commercial Cooking Equipment, NFPA 96, 1991

Roof Top Heliport Construction and Protection, NFPA 418, 1990

Safeguarding Building Construction, Alteration, and Demolition Operations, NFPA 241, 1996

Storage of Rubber Tires, NFPA 231D, 1998

Ventilation Control and Fire Protection of Commercial Cooling Operations, NFPA 96, 1998

Safeguarding Building Construction and Demolition Operations, NFPA 241, 1996.

Seattle-Tacoma International Airport Fire Department:
  Standard Operating Procedures

Aircraft Salvaging Operations

Airport Procedure Bulletin St. 4, Fire Department Inspections

Annual Report of Department Inspection

Code 19 Plan

Disaster Control Plan and Decontamination Teams

Inspection Report

Rules and Regulations

U.S. Department of the Air Force. *Technical Manual 100-25-172, Ground Servicing of Aircraft and Static Ground/Bonding,* December 1992.

GROUND HANDLING
REGIONAL JETS

CHAPTER 7

Aircraft arrivals
Wing walkers (guideperson)
Signaling
Chocking
Passenger handling
Ground support equipment
Aircraft departure
Pushback operations
Taxi-out operations

The basic safety tenets for ground handling regional jets are essentially the same as those for larger mainline aircraft; however, some differences exist. First, and foremost, regional jets are much smaller than mainline jets. They are lower to the ground, and space for loading/unloading is limited. This requires specialized servicing equipment. Secondly, regional jets cannot use most standard loading bridge facilities, thus off-gate parking is more common. Regional aircraft do use loading bridge facilities when the bridge is equipped with an appropriate regional jet adapter.

AIRCRAFT ARRIVALS

The ground crew charged with the responsibility of receiving and servicing the aircraft should have a briefing just prior to the aircraft's scheduled arrival. Topics to be discussed may include any unusual weather conditions, baggage/cargo loads, and safety reminders. Crews should conduct a pre-arrival foreign object debris (FOD) walk and visually check the aircraft parking area (i.e. safety diamond, circle, zone) to ensure all equipment is clear and will not make contact with arriving aircraft.

Wing walkers (guideperson)
Upon a flight's arrival in congested areas, wing walkers may be required. Prior to the aircraft approaching the parking area, wing walkers should conduct one last check to ensure the area is clear of any ground support equipment and FOD. Wing walkers should be in position and remain within sight of the guideperson as the aircraft is brought into the gate area.

Signaling
Upon arrival, before voice communication with the flight deck is established, hand signals are required to guide the aircraft into its final parking position. These signals must be understood clearly by the flight crew and should be in accordance with the International Civil Aviation Organization (ICAO) standards. Only trained and qualified personnel should be giving hand signals. Bright orange signal wands, or paddles, should be used for hand signaling. At night, illuminated wands are required.

Chocking
Upon arrival, the aircraft must be chocked in the front and back of nose or main gear per airline or manufacturer specifications. Chocks should be of

Figures 7-1 and 7-2. Wing walkers indicating the area is clear of traffic.

Photo courtesy of SkyWest Airlines

appropriate length and size to fully secure the nose wheel. If the aircraft will be parking overnight, both main gears should be chocked.

Passenger handling

Prior to deplaning, the passenger service agent should ensure no aircraft are running engines in the path of exiting passengers. Passengers should be advised by the flight crew or agent to exercise extreme caution when walking across the ramp and to follow the direction of passenger handling agents. Passengers may be exposed to high levels of noise, ground equipment, inclement weather and jet blast. In light of this, passengers should be closely supervised by passenger handling agents, if retrieval of carryon bags is permitted on the ramp. Passenger pathways from the aircraft to the terminal entrance must be clearly marked and fully illuminated during hours of darkness. Lines or stanchions to mark the boundaries of the pathway are recommended. Stairways leading into terminal buildings should be illuminated adequately, in good condition and equipped with hand rails.

Passengers with temporary or permanent disabilities may require special consideration. Hand-carrying passengers down the aircraft stairs creates the potential for injury to the passenger and/or employee. Specially designed ramps or lifting devices are required. (See photo below.) Employees should receive special training for handling passengers with temporary or permanent disabilities.

Ground support equipment

Extreme caution must be exercised when maneuvering motorized equipment in close proximity with the aircraft, due to the low profile and smaller size of the airplane. Wing tips and static dissipaters are particularly vulnerable to damage. Orange road cones, or delineator posts, should be placed slightly outside of the wing tips, aircraft nose and tail, to create an imaginary diamond shape around the aircraft. No motorized equipment should be allowed within these boundary lines except for those that have been authorized and are operated by trained employees.

Catering trucks designed for mainline aircraft are not recommended for use with regional jets. When positioning the catering truck to the aircraft a guideperson should be required. Once parked at the aircraft, the wheels should be chocked, the engine turned off,

Figure 7-3. Handicapped passenger being boarded via a portable ramp.

Photo courtesy of SkyWest Airlines

and the rear doors closed. Trucks should be equipped with extending safety rails on the loading platform. When catering activities are completed the truck should not be allowed to move away from the aircraft without the assistance of a guideperson.

Belt loaders designed specifically for use with regional jets or commuter aircraft should be considered for loading/unloading baggage. Positioning the belt loader to the aircraft must be done only when assisted by a guide person. The use of wheel chocks is recommended when the belt loader is in position at the aircraft.

Two potential hazards are associated with fueling regional jet aircraft: spillage and fire. Fueling should be performed only by trained and qualified personnel using fueling equipment in good operating condition. Fueling vehicles should be positioned outside of the aircraft safety zone and should have a clear path for rapid removal in the event of an emergency.

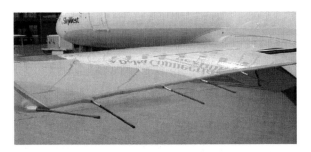

Figures 7-4 and 7-5. Static wicks on trailing edges of wings.

Photo courtesy of SkyWest Airlines

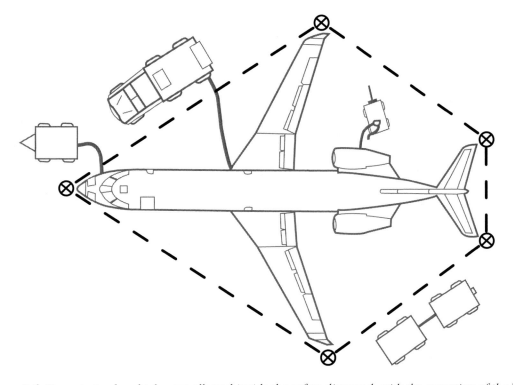

Figure 7-6. No motorized vechicles are allowed inside the safety diamond, with the exception of the belt loader.

Photo courtesy of SkyWest Airlines

NOTE: SPECIAL CONSIDERATION SHOULD BE GIVEN TO PLACEMENT OF FUELING VEHICLES DUE TO POTENTIALLY HIGH TEMPERATURE DISCHARGE OF APUs.

The fueling vehicle should be chocked. Proper grounding and bonding sequence should be followed to preclude the dangerous build-up of static electricity. Fire extinguishers should be of the appropriate size and in a proper location as specified in NFPA. Fueling with passengers on board the aircraft is not encouraged; however, if it is necessary, the aircraft main cabin door must be opened and a crew member should be on board with immediate means of egress available.

Aircraft deicing requires specialized training in accordance with government requirements. Only approved deice fluids must be applied to aircraft surfaces. Deicing equipment should be in good working condition and operated by qualified employees. Deicing vehicles that have an elevated boom with an aerial basket must have a means of voice communication between the basket and the driver in the vehicle's cab. When applying deicer fluid, consideration should be given to the direction of the prevailing winds and other employees working in the immediate vicinity. Local environmental regulations must also be taken into consideration with regard to fluid runoff and disposal.

AIRCRAFT DEPARTURE

The crew should have a pre-departure briefing just prior to pushback or engine start to ensure all concerned in the operation have the necessary information for a smooth departure. Everyone must understand their particular task and how it relates to other team members.

Pushback operations

If the departure requires pushback from the gate area, only trained employees should be allowed to operate the pushback tractor. Furthermore, the driver should be in constant communication with the flight deck. A straight back push can be accomplished with properly equipped wing walkers at both wings and, if necessary, an observer at the tail. A pushback with a turn, or in congested areas, should have an additional observer. All observers must maintain direct

Figure 7-7. Belt loader parked correctly inside the safety diamond.

Photo courtesy of SkyWest Airlines

Figure 7-8. Pushback being performed.

Photo courtesy of SkyWest Airlines

visual contact with the push back tractor driver. No one must be allowed in the vicinity of the nose gear or path of travel of the tractor.

Taxi-out operations

At locations where the aircraft departs from the gate under engine power (no pushback), care must be taken to avoid the hazards of jet blast. All equipment, especially rolling stock, must be adequately secured by means of brakes or chocks, to prevent being blown about by jet blast. Also, inclement weather can result in rain or snow being blown about the ramp. Clear and concise hand signals must be conveyed to the flight deck for taxi guidance. Wing walkers will stay with the aircraft until it has cleared the gate area safely.

WORKING GROUP MEMBERS

Chairman:

Jack Rentink – Skywest Airlines

Members:

William R. Jaggi – American Airlines

Thomas Jassoy

Larry T. Netherland – Delta Airlines

Joseph H. Nugent

Roland (Jim) Page – Page Safety Solutions

Dieter Pietsch – HDP Safety Solutions

Carlos Romero – LSG Sky Chefs

Fred Rose – Air Wisconsin

AIRCRAFT FUEL CELL
ENTRY SAFETY

CHAPTER 8

Modern transport aircraft carry jet fuel inside of tanks located within the airframe structure, which are known as integral fuel cells or, simply, fuel cells. Fuel cells must carry enough fuel to power the airplane for its maximum range, and they also house many other components of the fuel system. Like all parts of the airplane, fuel cells and the equipment located inside of them need periodic inspection, maintenance, and repair. Often these tasks require that aircraft maintenance personnel physically enter fuel cells in order to perform work.

Fuel cell entry work is one of the most challenging assignments in aircraft maintenance. Fuel cells are difficult to get in and out of, and there is little room to move around once inside. Many potential hazards confront the entrant, including the potential for a flammable and/or toxic atmosphere due to the presence of residual jet fuel and other chemicals. Identifying and controlling these hazards is essential to the safe completion of any fuel cell entry task.

This chapter is intended to provide safety and health guidance for entry into integral fuel cells aboard commercial jet transport airplanes for inspection, maintenance, and repair. Activities not covered include those associated with:

- Aircraft using other than commercial jet fuel
- Bladder or foamed fuel cells
- Assembly of new aircraft
- Entry into airplane spaces other than fuel cells, such as equipment bays

- Entry into fuel containing spaces not aboard aircraft, such as storage tanks or delivery trucks

Employers conducting fuel cell entry must develop and document their own safe entry procedures. While this chapter provides technical guidance that may be useful in developing such procedures, it is not intended to replace them.

Flight safety issues: Airplane fuel systems are flight safety critical. While this chapter contains information regarding the preservation of fuel tank integrity, flight safety issues relating to fuel system service are not comprehensively addressed. Always follow manufacturers' and regulatory requirements when servicing aircraft fuel systems.

Regulatory compliance: The activities described in this chapter are subject to numerous regulations from different government agencies that vary from region to region. Agencies that may have jurisdiction include those that regulate commercial aviation certification and operation, occupational safety and health, environmental protection, fire protection, and others. It is the employer's responsibility to ensure compliance with applicable local regulations. Where the content of this chapter conflicts with applicable regulations, priority must be given to the regulation.

Hazard assessment

Prior to making entry into an aircraft fuel cell, a thorough assessment of the potential hazards that may be encountered must be conducted. Many of these hazards, such as the difficulty in accessing the space and the presence of jet fuel, are inherent to fuel cells. Other hazards are created by the activities conducted inside by the entrant, including potential exposure to toxic and/or flammable materials used for cleaning, sealing and coating. Still others may result from activities occurring outside of and unrelated to the fuel cell. Regardless of how hazards may arise, identifying their potential and providing effective controls is essential before entry begins.

Hazards associated with fuel cell structure

Internal structure of fuel cells: The main fuel tanks of commercial airplanes are located within the wing structure, including the center tank embedded in the lower part of the fuselage (Figure 8-1). The primary design criteria for wings are that they are shaped to provide lift and structured to carry the load of the entire airplane in flight. Unfortunately, these criteria define a structure far from ideal for workers who must go inside. Wings are difficult to get into and out of, typically accessed by oval-shaped doors just large enough to admit a person with arms extended overhead (Figure 8-2).

Once inside, entrants are confronted with additional structure including ribs, spars and bulkheads that divide the wing into a number of smaller spaces called bays. Some bays do not have portals leading directly in from the outside, but instead must be accessed from adjoining bays through openings in intervening ribs and bulkheads. Bays located in the center tank and inboard wing may be large enough to stand up inside of, while outboard wing bays become progressively smaller as the wing tapers toward the tip, where only the worker's hands and arms may fit inside. Additional auxiliary fuel tanks may be located in the horizontal stabilizer or cargo holds.

Regardless of where fuel cells are located in the air-plane, their limited access and complex architecture makes them difficult to move around in, complicated to ventilate and challenging to communicate through.

Physical safety hazards: The tight confines of fuel cells contribute to a number of physical safety hazards. Most fuel cells have low overhead clearance inside, creating the possibility of head strike injuries. The presence of plumbing components and equipment brackets adds to this potential, and also may result in lacerations. The complex interior structure of fuel cells creates the possibility of the entrant being caught or hung up inside. Some airplane systems, such as the B-727 wing leading edge adjacent to the spar entry port, may injure workers if they are activated during access and so must be blocked and tagged prior to entry. Fuel cells may be entered while airplanes are jacked, but the airplane height should never be changed, either by jacking or by landing gear strut pressure adjustment while fuel cells are occupied. Most fuel cells are entered through access ports in the lower panel of the wing or center tank. Depending on the airplane model, access may require elevating the entrant to significant height, creating the potential for a fall.

Ergonomic hazards: Serious ergonomic stresses are inherent in fuel cell work. The tight confines and unforgiving structure of fuel cells forces workers to assume awkward postures in order to access all areas (Figure 8-3). This creates the potential for spinal and

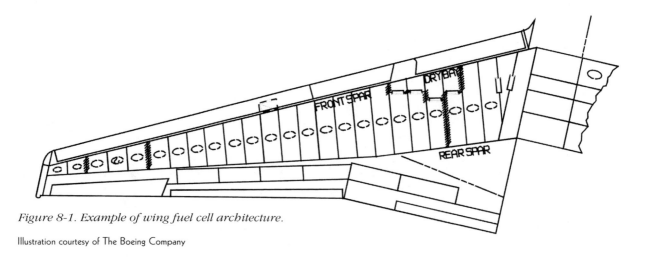

Figure 8-1. Example of wing fuel cell architecture.

Illustration courtesy of The Boeing Company

Figure 8-2. Fuel cell entry.

Photo courtesy of The Boeing Company

other musculoskeletal injuries. Entrants often must lay or kneel on the wing lower panel, a solid aluminum skin traversed at intervals by 2-inch (5 cm) high stringers. These structures can impart mechanical stress to entrant's muscles, tendons, and joints; and this pressure also may inhibit blood flow and normal nerve function. Pads and platforms taken into fuel cells may alleviate some of these concerns, but often are banned due to fire load and foreign object debris (FOD) issues. Cleaning, sealing, and fastening jobs may involve highly repetitive motions, and riveting may impart hazardous vibration to the worker's hands and arms.

Hazards associated with jet fuel

Jet fuel composition: Some residual jet fuel usually is present in airplane fuel cells. The most common commercial jet fuel is Jet Fuel A, or simply Jet A. Jet A primarily is a mixture of middle range petroleum distillates similar to kerosene. Jet A also may include varying low levels of other more complex petroleum hydrocarbons, and may also include various additives such as antioxidants, metal deactivators, static dissipaters, corrosion inhibitors, thermal stabilizers, biocides and fuel system icing inhibitors. While these other compounds may have unique hazards, they usually are of such low concentration that exposure potential is limited. The military fuels known as JP-4, JP-5, and JP-8 have varying hydrocarbon composition and additives that may present

additional hazards, but seldom are used in commercial transports.

Jet fuel combustibility hazards: Jet fuel is a combustible liquid, which means it can generate vapor that may burn or even explode. Even though jet fuel evaporates at a relatively slow rate at normal temperatures, vapor concentrations can rise to hazardous levels in fuel cells due to their small interior volume and limited ventilation. This effect is accelerated as temperature increases. Combustion of vapors depends on having the right ratio of fuel vapor to air. If the mixture is either too lean (not enough fuel vapor) or too rich (too much fuel vapor), combustion cannot occur. The range of vapor concentration in air where combustion can occur is called the explosive range (Figure 8-4). The lower (lean) end of this range is called the lower explosive limit (LEL), while the upper (rich) end is called the upper explosive limit (UEL). These concentrations usually are expressed in terms of volume percent, and vary from one material to another (Figure 8-5).

The primary means of controlling fire hazard during fuel cell entry is to dilute fuel vapors with air to the

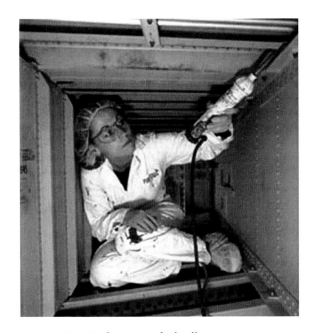

Figure 8-3. Working in a fuel cell.

Photo courtesy of The Boeing Company

point that they cannot burn. So, if an auxiliary ventilation system is used to introduce fresh air into the cell at sufficient volume, the vapors can be diluted to less than the LEL. Because vapor concentrations can change rapidly, a large safety factor is appropriate, so the limit commonly used for fuel cell entry is 10 percent of the LEL.

Jet fuel contact hazards: Skin contact with jet fuel may lead to drying and irritation. More severe effects – including dermatitis, rash and skin damage – may result from long term, repeated contact; studies on laboratory animals have found that skin cancer can develop in severely damaged areas. Eye contact may result in moderate irritation. Skin and eye contact must be avoided through correct work practices and the use of personal protective equipment.

Jet fuel inhalation hazards: Jet fuel can be inhaled in either of two phases: vapor or aerosol. The inhalation hazard most likely to be faced by a fuel cell entrant is jet fuel vapor. The vapor phase is composed of airborne hydrocarbon molecules that have evaporated from liquid fuel. Inhalation of excessive concentrations of jet fuel vapor may cause irritation of the respiratory system, as well as short-term central nervous system effects including headache, dizziness, nausea, impaired coordination and fatigue.

Jet fuel also may become airborne as an aerosol or mist. A common situation where jet fuel aerosol can occur is in the exhaust stream of a jet engine being cold-started. Inhalation of excessive concentrations of aerosolized jet fuel can cause the same symptoms as those listed above for jet fuel vapor as well as other, more serious, health consequences. However, exposure to jet fuel aerosol is unlikely while working inside of fuel cells on commercial airplanes. The primary means of preventing inhalation of excessive concentrations of jet fuel vapor is dilution with fresh air via an auxiliary ventilation system. Where ventilation cannot reliably control this hazard, respiratory protection may be required.

Jet fuel exposure limits: Many hazardous chemicals have exposure limits, which are published

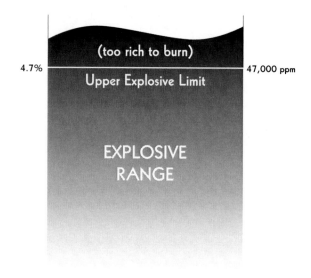

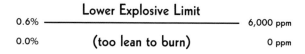

Figure 8-4. Jet fuel explosive range at ambient temperature and pressure.

maximum air concentrations that should not be exceeded for workers without respiratory protection. These limits may come from regulatory agencies or advisory organizations, and are based on the potential inhalation health effects of the chemical in question. Many exposure limits have been suggested for jet fuel. One such limit is the threshold limit value (TLV) for kerosene/jet fuels published by the American Conference of Governmental Industrial Hygienists (ACGIH). The TLV is a health-based advisory standard that may be used as a guideline for limiting exposure when no appropriate regulatory standard exists. Always check with your local occupational safety and health jurisdiction for an applicable exposure limit for jet fuel.

Additional atmospheric hazards

Oxygen concentration: Oxygen deficiency or enrichment is another potential atmospheric hazard associated with fuel cell entry. Ambient air normally contains approximately 21 percent oxygen. If the

Chemical	LEL	UEL	Source
Jet A fuel	0.6%	4.7%	Chevron MSDS for Jet Fuels 11/18/97
Methyl ethyl ketone (MEK)	1.8%	11.5%	DuPont MSDS for MEK 5/23/00
Methyl propyl ketone (MPK)	1.6%	8.7%	Eastman MSDS for MPK 12/15/99
d-Limonine	0.6%	7.0%	Inland Technology MSDS for Citra-Safe 8/3/98

Figure 8-5. Explosive range concentrations for common fuel cell chemicals.

% Oxygen	Effect
20.9	Normal
19.5	Safety limit
17	First sign of hypoxia; some increase in breathing volume and accelerated heart rate
16	Impaired judgment and breathing
14	Increased breathing volume and heart rate; impaired attention, thinking and coordination
12.5	Immediately dangerous to life and health (IDLH)
10	Very faulty judgment and muscular coordination; Intermittent respiration
6	Spasmatic breathing, convulsive movements; death in minutes

Figure 8-6. Physiological effects of oxygen deficiency.

oxygen in the air is diluted to a significantly lower concentration – perhaps by the introduction of an inert gas such as nitrogen into a fuel cell – it may reach the point where entrants become incapacitated or even asphyxiated (Figure 8-6). Inert gases such as nitrogen or argon should never be used in fuel cells during entry unless special procedures are followed. Fuel cells should be ventilated during entry to prevent oxygen depletion by the entrant's respiration. Oxygen enrichment also is a hazard because a high oxygen atmosphere supports rapid combustion of materials that may not otherwise burn. Gaseous oxygen never should be used in a fuel cell for any reason.

Other hazardous materials used in fuel cells: Many of the ingredients in the solvents, sealants and coatings used in fuel cells are toxic and may be subject to worker exposure limits. They also may evolve flammable vapor. Some of these chemicals actually may have greater hazard potential than jet fuel. Information about these products and the necessary precautions for their use may be found in the Material Safety Data Sheet (MSDS) provided by the manufacturer. Another source of hazardous air contaminants generated inside of fuel cells are particulates that may be created by sanding or grinding on fuel cell sealants and coatings that may contain chromates or other heavy metals. The potential hazards of all materials used or worked on inside of fuel cells must be identified and controlled.

Hazardous materials used outside of fuel cells: Chemical hazards from activities outside of the fuel cell also must be considered. Certain activities such as spray painting, corrosion inhibitor application, welding, sanding or grinding may produce hazardous air contaminants that may be introduced into the fuel cell, potentially exposing the workers inside. This is especially true if an exhaust system is used alone to ventilate the fuel cell, or if the hazard generating process is conducted next to the intake for a supply ventilation system. Another important outside source of hazardous air contaminants are the internal combustion engines in vehicles like tugs or boom lifts, which may generate carbon monoxide and other toxic by-products. Processes and equipment that generate hazardous air contaminants should be excluded from the fuel cell entry work zone and the area around the supply ventilation intake.

Other hazards

Ignition sources: Even where flammable vapors are diluted via auxiliary ventilation, potential ignition sources must be eliminated from fuel cells and surrounding areas to further reduce fire or explosion hazard potential. All electrical components installed in the fuel cell, such as boost pumps and fuel level sensors, must have circuit breakers pulled and tagged during entry. Electrical equipment brought into fuel cells such as work lights, flashlights, and test equipment must be certified as acceptable for use in a hazardous environment.

Another important ignition source is static electrical discharge. All textiles brought into the fuel cell such as worker's garments, ventilation ducts, and plastic bags must be made of a conductive material to prevent the buildup of static charges. Fuel cell entry work should not be conducted outdoors during periods of lightning. Hot work such as grinding or breaking rivets can be conducted only under special conditions as specified by a qualified person familiar with aircraft fuel cell fire issues.

Physical exposure hazards: Exposure to a number of physical agents also is possible during fuel cell entry. Excessive noise is common, created by ventilation systems or operations such as riveting. Check with local regulations regarding noise exposure limits and the requirements of hearing conservation programs. Entrants also may be exposed to excessive heat or cold. This is especially common where work must be conducted in extreme climates, and is complicated by the use of protective garments such as fuel cell coveralls.

Some fuel measuring system densitometers emit low levels of radiation. Fuel cell entrants should not remain near densitometers for extended periods of time. Entrants also are subject to radiation exposure if inspection X-ray equipment is used near an occupied fuel cell without proper controls. The hazards of exposure to lower or higher than ambient atmospheric pressure also must be considered for operations such as negative pressure leak testing or when using high flow ventilation systems.

Entrant factors: Certain physical conditions are risk factors for personnel entering fuel cells. Due to the difficulty in rescuing a stricken fuel cell entrant and the potential for chemical exposure, employees with heart or respiratory conditions, nervous system conditions including seizure disorders, metabolic disorders, abnormal liver function, claustrophobia, or chronic skin disorders may be at greater risk. Due to ergonomic stresses inherent in fuel cell work, employees with a history of back injuries or other musculoskeletal disorders may be prone to further injury. Employees who regularly perform fuel cell work should receive preassignment and periodic health evaluations.

Multiple employer entries: Coordination of fuel cell entry activity is essential when more than one employer is involved, such as when the operator's maintenance crew and a fuel cell repair contractor must work together. Each employer should have a complete fuel cell entry program, and must take responsibility for its own activities. The host employer should inform the contractor of all known potential hazards, and of the control requirements the host typically uses. The contractor must inform the host employer of any known new hazards discovered or created during entry. All must cooperate to ensure safe conditions are maintained throughout the operation.

VENTILATION

Fuel cells must be ventilated to control fire and toxic inhalation hazards during entry. Mechanical ventilation systems must be configured to ensure an adequate supply of fresh air is provided to all areas entered by maintenance personnel. However, the tight confines and complex architecture of internal wing structures make them particularly difficult to ventilate effectively. The nature of the process to be conducted inside of the cell is another important factor. Air monitoring for fuel vapor and other air contaminants typically provides the best indication of ventilation adequacy.

Supply versus exhaust: Supply (blower) or exhaust systems, or a combination of the two, may be used for fuel cell ventilation (Figure 8-7). Blowers equipped with flexible ducts are usually the system of choice, as they are typically more effective for several reasons. Blower systems "throw" air in a uniform direction a considerable distance away from the end of the duct, allowing the stream to be directed into areas where air is needed most. Exhaust ducts draw more or less equally from every direction, which limits the air velocity in any one direction as the distance away from the end of the duct increases. Blower systems are inherently more fire safe because only clean outside air is drawn through the fan section, whereas air that may contain flammable vapors is drawn through all parts of an exhaust system. Filtered blower systems can minimize fuel cell contamination by airborne debris, while exhaust systems draw through all openings into the cell, which may allow entry of dirt or other debris.

Blower and exhaust systems may be used quite effectively together, creating a "push-pull" system within the wing. This arrangement has all the advantages of a blower system, and also allows the exhausted air to be removed from the hangar or filtered to avoid buildup of fuel and/or solvent odors in the work area.

Types of air movers: A variety of fuel cell ventilation equipment is available. Fan systems may be used for blowing or exhausting, and are available as portable units (Figure 8-8) or may be provided as fixed equipment in dedicated maintenance hangars. Portable fans may be powered by electricity or compressed air. All fan systems used must be acceptable for use in a flammable/combustible environment, such as those having sealed electrical systems, nonsparking fans, and conductive ductwork to dissipate static electrical charges.

Another option for fuel cell ventilation is the venturi, which uses compressed air to induce airflow through

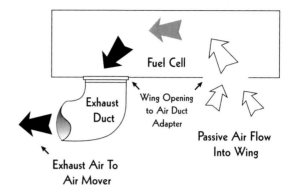

Exhaust Ventilation

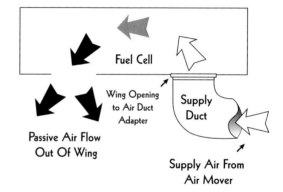

Supply Ventilation

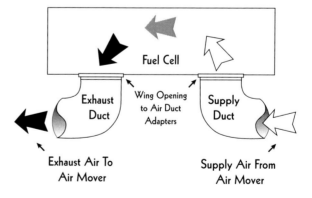

Push-Pull Ventilation

Figure 8-7. Types of fuel cell ventilation.

Illustration courtesy of Bryan Bauer, CIH

a tube (Figure 8-9). Venturis often are used in exhaust mode for achieving the desired number of air changes after emptying the tank and prior to entry. They may be attached to fuel cell doors via adapters, or directly to over-wing fill ports on older airplanes.

Other factors to consider when selecting air movers for purchase are physical size and portability, power requirements, noise, operating temperature, and heat generation considerations. Many other features are available for tank ventilation systems, including failure alarms, air conditioning for blowers and charcoal filtering units for exhaust systems.

Sizing air movers: The size of the space and the activity to be conducted inside are primary determinants when sizing air movers. A single low-volume blower may be sufficient for a single wing bay, while larger or multiple blowers may be required for larger center tank bays. Likewise, a single low-volume blower may be sufficient to dilute fuel vapors in a mostly dry tank, but additional or larger blowers may be required to dilute vapor to a safe level during operations such as solvent cleaning or painting. It is important to remember that any restriction to flow such as kinked ductwork or dirty inlet filters will reduce the amount of air delivered.

When sizing new equipment for purchase, make sure the fan is capable of generating the needed air flow at

Figure 8-9. Venturi.

Photo courtesy of American Airlines

a level of resistance (or "static pressure") representative of in-service conditions. The services of a qualified ventilation engineer or industrial hygienist may be needed to properly size and select fuel cell ventilation equipment.

Ventilation ducts: Flexible ductwork for fuel cell ventilation usually consists of two types: helical and collapsible. Helical flex duct has a stiff spiral wire winding through it to hold its tubular shape. This kind of duct has the advantage of maintaining an open passage for air flow unless radically kinked, but may be stiff and hard to handle, and creates a partial blockage through entry doors and bulkhead passages. Collapsible duct is made from poly sheeting like a heavy-duty plastic bag. It collapses to allow free movement for the entrant, but when collapsed or kinked airflow is severely reduced. Regardless of which kind of flexible ducting is selected, it must be:

Figure 8-8. Portable air blower.

Photo courtesy of The Boeing Company

• Made of a fire retardant, chemical resistant material
• Conductive to dissipate static electrical charges (applies to connectors/adapters as well)

- Sized correctly, usually in the 6- to 10-inch (15 to 25 cm) diameter range, to accommodate the needed air flow
- Frequently inspected and replaced when punctured or otherwise damaged

Ducts must be long enough to reach the affected area within the wing, and also allow the blower to be located far enough from the entry point to ensure contaminated air will not be recirculated. It should be noted, however, that resistance to flow through a duct increases with length.

Pre-entry and during-entry ventilation: As feasible, fuel cells should be ventilated continuously from the time they are opened until they are closed again. Ventilation typically takes place in two phases: pre-entry and during-entry. Pre-entry ventilation (or purging) is conducted after the tank is drained to prepare the space for entry. Goals for pre-entry ventilation include evaporating as much fuel residue as possible from surfaces within the tank and diluting fuel vapor concentrations to fire safe and – if possible – health safe levels. Pre entry ventilation often is accomplished using venturis, which are allowed to run as required to achieve many air changes within the wing before pre-entry air monitoring is conducted. During-entry ventilation usually requires more focused air flow because it must ensure safe concentrations are maintained in the immediate vicinity of the entrant, even if additional airborne contaminants are generated by operations such as solvent cleaning or painting. During-entry ventilation usually is accomplished using ducted blowers or push-pull systems. Appropriate steps must be taken, such as signage and attendant checks, to ensure that ventilation is not interrupted during entry. Ventilation failure alarms also are available. Entrants should evacuate the cell immediately if ventilation fails.

During-entry ventilation configuration:

The placement of ventilation system components within wing structures requires thorough knowledge of wing architecture and careful planning to ensure effective air delivery (Figure 8-10). For smaller wing bays, supply ducts typically are placed in one of the two bays on either side of the bay to be entered (Figure 8-11). This allows air to flow into the occupied bay through the fuel flow openings in the intervening wing rib. Of course, air cannot flow through solid bulkheads at tank ends or dry bays, and the potential for the entrant's body to block airflow also must be considered. As the diagram shows (Figure 8-10), selective installation of fuel cell access doors or dust covers can help to enhance flow-through ventilation.

For larger inboard bays and center tanks, one or more ducts usually are placed in the occupied bay. Duct ends should be oriented to ventilate the occupied portion of the bay, or diffuser systems such as perforated ducts with blocked ends may be used to distribute air throughout the space. Avoid ventilation configurations where the only opening for passive airflow to leave or enter the cell is the access hole. This situation can create air pressure fluctuations while the entrant's body blocks the hole during entry and exit, and this pressure change may injure the entrant's ears.

AIR MONITORING

The atmosphere inside of fuel cells must be evaluated both before and during entry. Measuring for oxygen content and the airborne concentration of fuel vapor and other contaminants is essential to ensure safe entry conditions are maintained, and also to select the proper level of respiratory protection. Atmospheric monitoring equipment – properly selected, calibrated and used – is critical to ensuring the safety of the entrant. A number of different portable instruments capable of measuring airborne concentrations of jet fuel and other contaminants are available (Figure 8-12).

Combustible gas indicators: An instrument commonly used to measure percent LEL is the combustible gas indicator, or CGI. The typical CGI pumps the sample of air through a probe and into a sampling chamber. There, the sample flows over a hot wire or catalytic bead, and any combustible gas or vapor in the sample burns and changes the temperature of the

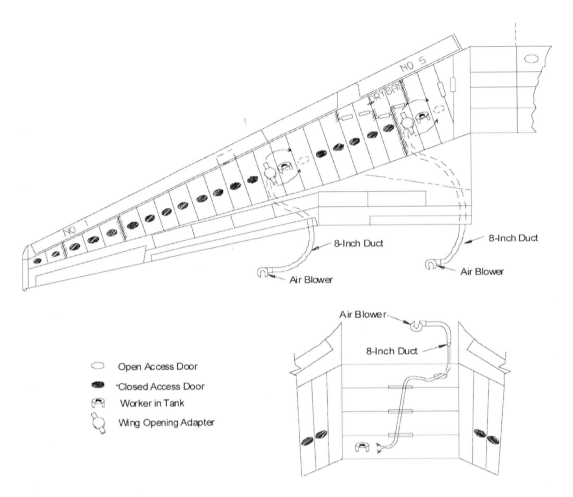

Figure 8-10. Example ventilation configurations for wing and center tank.

Illustration courtesy of The Boeing Company

wire. This change in temperature is accompanied by a change in resistance to a small current flowing through the wire, which can be measured and indicated as a concentration on the dial or digital display of the meter. CGIs have proven reliable over the years for percent of lower explosive limit (LEL) measurements. However, this method of detection lacks the precision necessary for the parts per million measurements needed to determine respiratory protection requirements. Further, because they actually burn the sample, CGIs can function only where there is adequate oxygen. Also, most CGIs do not direct-read jet fuel concentration; instead they require the user to multiply the result by a correction factor.

Photoionization detectors: Another detector available for fuel cell use is the photoionization detector, or PID. As the sample is pumped into the chamber, it is bombarded with ultraviolet energy. This energy ionizes the contaminant molecules, and their accumulated charge is measured and translated into a concentration for display. The PID offers the advantage of precision and accuracy at low concentrations, and sensitivity to a wide range of compounds even in an oxygen deficient atmosphere. PIDs are now available packaged with oxygen and LEL sensors in a single, small portable instrument. Some even have circuitry that allows them to direct read in parts per million of jet fuel vapor or common fuel cell solvents such as methyl ethyl ketone.

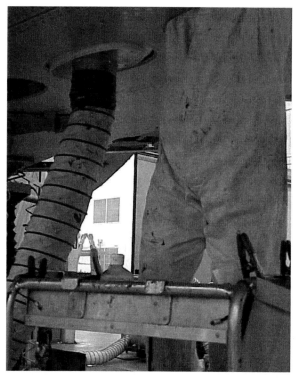

Figure 8-11. Duct placement in a wing fuel cell.

Photo courtesy of The Boeing Company

Instrument criteria: Regardless of which vapor detection system is employed, any device used for fuel cell atmospheric monitoring should be:

- Capable of indicating oxygen content, with resolution of 0.1 percent or less
- Capable of indicating percent of lower explosive limit, with resolution of 1 percent or less
- Capable of indicating parts per million of jet fuel or other hazardous vapor or gas as required, with resolution of 10 ppm or less if the instrument will be used to determine level of respiratory protection
- Able to indicate in real time, with minimal delay
- Equipped with displays that are easy to read, even in low light
- Equipped with adjustable alarms, which can be set at appropriate warning levels
- Have Underwriter's Laboratory (UL), Factory Mutual (FM), or equivalent certification as acceptable for use in hazardous locations

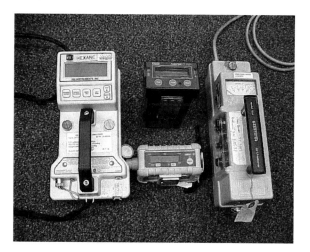

Figure 8-12. Air Monitoring Instruments.

Photo courtesy of The Boeing Company

- Small and light enough to be easily carried inside of fuel cells
- Rugged enough to withstand the shock of normal use, and resistant to fuel and solvents
- Able to be equipped with a pump and probe to allow remote sampling prior to entry
- Operable on battery power for at least 8 hours
- Able to be field calibrated easily
- Provided with clear operating instructions and conversion factors for jet fuel and other contaminants as required
- Capable of operating within expected temperature and pressure ranges
- Shielded from interference by radio signals

Pre-entry monitoring and during-entry monitoring: Fuel cell atmospheric monitoring usually is done in two phases: pre-entry and during-entry.

Pre-entry monitoring is conducted to ascertain the space is safe to enter, and to determine what level of respiratory protection is required (Figure 8-13).

During-entry atmospheric monitoring is conducted to verify safe conditions are maintained inside of the fuel cell throughout the entry. It is especially important to conduct monitoring during the actual entry if chemicals such as solvents or paints are being used inside,

Figure 8-13. Pre-entry air monitoring.

Photo courtesy of The Boeing Company

or if the effectiveness of ventilation at the entrant's location is in question. During-entry monitoring may be conducted by an employee outside the fuel cell, or by the entrant. Because it is most desirable to sample near the entrant's breathing zone, having the entrant carry the detector inside is often most effective. It is essential that the entrant carry the monitor when accessing remote bays such as the center tank aft section, or when there is no access from outside the cell, such as during negative pressure leak testing.

Monitoring procedures: Always follow the instrument manufacturer's instructions for air monitoring. However, this general procedure should be followed in order:

1. Check battery condition.
2. Calibrate meter per manufacturer instructions (Figure 8-14), or verify that calibration certification is valid.
3. Verify and record conversion factors for jet fuel vapor or other contaminant if required.
4. Ensure alarm limits meet local standards. The use of additional safety factors (e.g., setting the alarm to go off before the limit is reached) may be appropriate for some situations.
5. Ensure the fuel cell is prepared for entry including draining, sumping and ventilating.
6. Ensure the percent oxygen reads ~20.9 and LEL and ppm vapor read zero in fresh air just prior to

measuring the space. Record results.
7. Take and record measurements inside of the fuel cell opening. All measurements should be taken in this order:

   - Oxygen content
   - Percent LEL of fuel vapor
   - ppm fuel vapor, if required

8. Sample all areas that the entrant will access as feasible. If using a remote probe and pump, be sure to allow sufficient time for the pump to move the sample through the remote probe, per manufacturer's instructions. Record results.
9. Take additional readings near the entrants breathing zone during the entry at appropriate intervals, and record. Always take additional readings whenever a change in vapor concentration may occur, such as when a new chemical process begins or when ventilation ducts are moved. Continuous monitoring also is an option.

Oxygen readings: Special attention must be paid to oxygen readings in fuel cells. The range of readings commonly accepted as safe are 19.5 to 23.5 percent oxygen, and these limits are typically used as instrument alarm settings. However, under normal conditions the oxygen content of a properly ventilated airplane fuel cell should not vary much from the ambient level. Therefore, any significant variation from 20.9 percent oxygen should be investigated immediately. Oxygen in fuel cells may be reduced to dangerous levels by inert gases such as nitrogen or argon that may have been used deliberately to inert the fuel tank, or inadvertently introduced to drive pneumatic tools or to pressure test fuel plumbing systems.

Conversely, if the oxygen is being diluted by a toxic gas or vapor, levels of that contaminant will become very high before oxygen levels show a significant reduction. Another consideration is work conducted at high altitudes or at reduced pressure. While the meter may still correctly read 20.9 percent oxygen, the partial pressure of oxygen already may be reduced to the extent that any further reduction could be hazardous. A trained safety and health professional should be consulted to establish safe oxygen levels for high altitude or reduced pressure entries.

LEL readings: The commonly used "fire safe" level for jet fuel vapor is equal to 10 percent of its lower explosive limit. The percent LEL alarm should be set at this level, and entry shall not be allowed if it is exceeded. If the instrument used requires a conversion factor, the alarm point should be set to go off at the reading equivalent to the true 10 percent LEL concentration for the fuel or other contaminant being measured.

ppm readings: The "health safe" level for jet fuel – the level at which entrants may enter without respiratory protection – may be based on the TLV or on local standards, which may vary from one jurisdiction to another. The ppm alarm should be set at this level if entry without respirators will be allowed. When respirators are worn, the acceptable concentration is determined by multiplying the appropriate exposure limit times the protection factor for the respirator (see respiratory protection section). Note that ppm readings may not be required where respirators with very high protection factors (such as air supplying respirators) are mandatory. It also should be noted that some respirators can provide protection from jet fuel vapor concentrations in excess of 600 ppm, which is equivalent to 10 percent LEL.

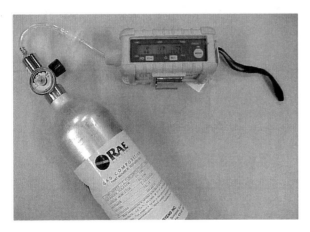

Figure 8-14. Instrument calibration.

Photo courtesy of The Boeing Company

Although the entrant may be protected from inhalation of vapor, entry must still be denied for fire safety reasons.

Monitoring other contaminants: Other chemicals used in fuel cells, such as solvents or coatings, may result in hazardous conditions that require monitoring. The solvents typically used for cleaning and some of the ingredients in sealants and coatings can be detected by CGIs and PIDs. However, these instruments cannot discriminate between different contaminants. When fuel vapor levels are low, accurate readings may be obtained for a single contaminant such as methyl ethyl ketone if the appropriate conversion factor is used. However, if mixed vapors are present, the instrument will read out a combined concentration that is difficult to interpret. Methods for monitoring of chemicals other than jet fuel, especially where multiple contaminants are present, should be developed by a safety and health professional trained in air monitoring techniques.

PERSONAL PROTECTIVE EQUIPMENT (PPE)

A variety of PPE is available to facilitate safe entry into aircraft fuel cells (Figure 8-15). Protection must address all potential chemical exposure routes for fuel cell entrants: skin contact, eye contact and inhalation. Hearing protection also may be required.

Hand protection: Skin contact is most likely to occur on the entrant's hands. Chemical protective gloves generally are required for any entry into previously fueled airplanes. Where the only potential contact is with residual fuel, nitrile gloves offer good protection. For other chemicals such as solvents, sealants or paints, consult glove manufacturer's recommendations. When choosing glove types, the breakthrough time for the specific chemical challenge is an important factor. It is desirable to select a glove that will protect for the full duration of the process, because fuel cell entrants are reluctant to exit just to change gloves, and spare gloves taken inside accidentally may be left there. Other factors to consider include glove thickness and flexibility, because of the dexterity required for fine work such as handling small

Figure 8-15. Fuel cell personal protective equipment.

Photo courtesy of American Airlines

fasteners. Gloves should be discarded whenever they show signs of degradation such as blistering, cracking or swelling.

Coveralls: Fuel cell entrants must wear protective garments – which may include coveralls, head coverings and boot socks – to protect both themselves and the airplane. These garments must be static-dissipating to avoid creating an ignition source inside of the fuel cell; for this reason heavy cotton fabric usually is chosen. While garments made of cotton may offer some protection against penetration by heavy sealants or paints, they easily are soaked through by solvents or fuel. Good work practices such as thorough drying of fuel cells and using the minimum amount of solvent necessary to complete a cleaning task are essential in avoiding excessive skin contact. If body contact with fuel or other liquid chemicals is possible, fuel-resistant coveralls are available. It should be noted, however, that coveralls made of liquid-tight materials may contribute to heat stress for the entrant because evaporative cooling from sweat is restricted.

Eye protection: Safety glasses with side shields provide adequate eye protection for many fuel cell entry jobs. However, where chemicals may be used overhead, chemical splash goggles should be considered because they create a seal around the eyes that will keep out any chemicals that run onto the

face. Eye protection also is provided by the use of a full facepiece respirator.

Respiratory protection: Respiratory protection is required whenever the potential exists for exposure limits to be exceeded. Ventilation and correct work practices always should be the primary means of controlling toxic atmospheres. However, where ventilation systems and best practice procedures cannot limit vapor levels, a respirator may be required. The two general classes of air-supplying and air-purifying respirators are available, each having advantages and disadvantages. Respirator selection must be conducted by qualified safety and health professionals. All respirator users must be medically qualified, trained and fit tested before using respirators in a hazardous environment. Supervisors of respirator wearers also should be trained.

Respirator protection factors: Different types of respirators offer varying degrees of protection from hazardous atmospheres. To quantify these differences, numerical protection factors are assigned to each type. The protection factor number is multiplied by the exposure limit for the chemical at issue, and the product of this calculation is the maximum use concentration for that combination of chemical and respirator. The protection factor for a given type of respirator can be found in the manufacturer's literature. In general, full facepiece respirators offer higher protection factors than half-masks, and air-supplying respirators offer higher protection factors than air purifying. It should be noted that use of some respirators provides protection from jet fuel vapor concentrations in excess of 600 ppm, which is equivalent to 10 percent LEL. Even though the entrant may be protected from inhalation of vapor, entry must still be denied for fire safety reasons.

Air-supplying respirators: Air-supplying respirators used in fuel cells are typically of the continuous flow or pressure-demand air line type (Figure 8-16). They rely on having a clean source of compressed air from outside the airplane, which is fed through a hose attached to a tight-fitting half-mask or full facepiece, or a loose-fitting lightweight hood. Because

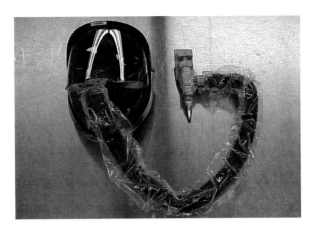

Figure 8-16. Air-supplying respirator.

Photo courtesy of The Boeing Company

Figure 8-17. Air-purifying respirators.

Photo courtesy of The Boeing Company

full facepiece air-supplying respirators offer the highest level of protection, they often are specified for fuel cell entries where high exposures are expected. They also have the advantage of the air stream's ability to cool the entrant and prevent fogging of the full facepiece visor. A major disadvantage is the difficulty of trailing an air hose through the confines of fuel cells. It should be noted that the loose-fitting, lightweight air-supplying hood is the only respirator that can be used by entrants with beards or other facial features that interfere with the seal of a tight-fitting respirator.

Air-purifying respirators: Air-purifying respirators typically are of the dual chemical cartridge type (Figure 8-17). For jet fuel and cleaning solvents, as well as the solvents present in most sealants and paints, the organic vapor cartridge is used. If spray painting or other particulate-generating operations are conducted, a pre-filter may need to be added. In any case, a thorough evaluation of the ingredients listed on the MSDSs for all products used is essential to ensure correct respirator selection. Because they operate at negative pressures, the protection factors for air-purifying respirators are not as high as those for air-supplying respirators with similar facepieces. However, for low-exposure entries, many entrants prefer air-purifying respirators for the convenience of a compact self-contained respirator with no hoses or other attachments.

There are several critical limitations on the use of air-purifying respirators. Most important, air-purifying respirators cannot be used in oxygen-deficient atmospheres. They also cannot be used in atmospheres with chemical concentrations that exceed "immediately dangerous to life and health" (IDLH) values. Because they operate at negative pressure, air-purifying respirators cannot be used by entrants with facial hair or other features that interfere with the facepiece seal. Chemical cartridges have a finite capability to sorb contaminant, like water in a sponge. Therefore, maximum use times must be calculated for each chemical that will be encountered. Respirator manufacturers offer assistance in determining changeout schedules for their products.

Hearing protection: Ear plugs or muffs may be required to protect an entrant's hearing from noise sources such as high-volume ventilation systems and riveting. Plugs usually are preferred because they are less likely to create interference during entry. However, some riveting jobs may require both plugs and muffs to achieve the necessary sound level reduction. Noise monitoring by a trained professional may be required to determine the appropriate level of hearing protection needed.

Entry team roles

Fuel cell entry is a cooperative activity. Employees must work together to fulfill a number of specific roles to accomplish a safe and successful entry. Some of these roles may overlap, and a single employee may fulfill more than one role. But trained employees must be identified specifically to meet each of these essential functions: entrant, attendant, detector operator, authorizer and emergency responder. Each of these roles bears responsibility for the safety of the entrant, and any participant may terminate the operation if he or she believes entry is unsafe.

Entrant: The entrant is a specially trained employee who actually goes inside the fuel cell. Specific entrant responsibilities include wearing proper personal protective equipment, employing safe work practices, communicating with the attendant, and evacuating the fuel cell if conditions become unsafe or if symptoms of exposure occur. The only other role that the entrant may fulfill during a given entry is detector operator if atmospheric monitoring is conducted from inside the fuel cell.

Attendant: An attendant must be provided for all fuel cell entries (Figure 8-18). The attendant's essential duties include:

- Keeping account of, and monitoring the well-being of, all entrants
- Being alert to the development of hazardous conditions outside of the space
- Preventing unauthorized personnel from entering the space
- Directing entrants to exit the space if unsafe conditions occur or if symptoms of exposure are noted in the entrant
- Summoning and assisting rescue services as needed

Attendants must be trained in fuel cell entry safety, and typically receive the same training as entrants. Attendants are expressly prohibited from entering fuel cells during the routine course of their duties. The only exception to this rule is if the attendant also is designated, trained and equipped as a rescuer and is relieved by another attendant during an actual rescue. Attendants also may serve as detector operators and/or authorizers. For fuel cell entry, attendants may monitor a single entrant or all of the entrants on one

Figure 8-18. Fuel cell entry attendant.

Photo courtesy of American Airlines

airplane. Attendants may be assigned to monitor multiple spaces or even multiple airplanes, as long as all the responsibilities listed above can be met. Attending to multiple spaces in large aircraft maintenance facilities may require the application of an electronic monitoring system.

Detector operator: The detector operator is responsible for atmospheric monitoring. The entrant, attendant or authorizer also may serve as detector operator.

Authorizer: The authorizer is the individual identified by the employer who signifies that all safe entry requirements are met by signing the entry permit and checklist, and takes overall responsibility for the fuel cell entry. Authorizers usually are shop supervisors, leads or other senior employees with comprehensive fuel cell entry experience. Authorizers also may serve as attendants or detector operators.

Emergency responder: Emergency responder responsibilities will be covered in the emergency response section later in this chapter.

COMMUNICATION

Attendants and entrants must have the capability for constant communication. For simple one-on one deployment in small spaces such as wing fuel cells, visual verification from outside the space that the entrant is moving, or basic voice inquiries are sufficient. When larger and more complex inboard cells are entered, visual contact becomes impossible. Voice communication also becomes more difficult because of competing noise coming from ventilation systems or other nearby processes, and because the entrant and/or attendant may be wearing hearing protectors or may be hearing impaired. A planned communication system of tapping on airplane structure, or tugging on a lifeline or air hose may be established, although such systems are obviously limited as to the amount of information that they can convey. Check-in communication between attendant and entrant should be accomplished at agreed to regular intervals. Either party should have the ability to initiate communication at any time.

Electronic communication may be used for fuel cell entry. The simplest systems consist of a pair of two-way radios or hard-wired intercoms. Wireless systems usually are preferred because of the difficulties associated with pulling a wire through fuel cell structures. Electronic communication may allow a single attendant to monitor multiple entrants on more than one airplane in a single facility. Complete turnkey systems are available which allow entrants to electronically log in to specific locations; the computerized system will periodically poll each entrant in turn, and sound an alarm if no response is received. Any electronic communication components taken into fuel cells must be certified for use in flammable/combustible environments. They should be as small as practical for ease in carrying and placement; and should be equipped for tone, flasher and/or vibrate alert. Voice capability also is desirable.

TOOLS AND EQUIPMENT

Proper tools and equipment enable work inside fuel cells to be carried out in a safe manner. All fuel cell entry support equipment should be thoroughly inspected prior to use. Equipment requirements for ventilation, air monitoring, communication and personal protection have been covered in separate sections. A list of requirements for other fuel cell entry equipment follows:

Signage: A number of signs or other visible warnings should be in place during fuel cell entry. The first is posted at the perimeter surrounding the airplane and should state that fuel cell entry is in progress, with language such as:

Caution

Open Fuel Cells

The second sign is posted at any disconnected ground power or battery power connections. It reads:

Caution

Open Fuel Cells

Do Not Apply Power

A third sign is posted at the entry point, typically on the ladder or entry stand, reading:

Caution

Worker Inside

Move No Equipment

These signs should be removed when entry is complete and when fuel cells are closed.

Locks/blocks/tags: Appropriate tagged locks or blocks must be available for circuit breakers and moveable components that may create hazards during the entry.

Ladders: Stepladders may be used, but they must be of the correct class, in good repair, and properly sized to allow the entrant to get in and out of the fuel cell easily. Entrants must be especially cautious when exiting to a ladder because they typically cannot see their feet. It is helpful to have the attendant steady the ladder during entry and exit. When ladders are used on elevated platforms, they cannot be positioned such that the entrant could fall over the platform guardrails if the ladder should tip. Wooden or composite ladders are preferred because they do not need to be grounded.

Lifts: Powered scissor or articulating boom lifts are the preferred means of accessing extremely high locations, such as wing fuel cells on large airplanes where under-wing work stands are not provided. Due to the potential for flammable fuel vapor, all powered lifts used for this purpose must be rated for use in a flammable/combustible environment. Entrants must be trained in safe lift operation, and may not climb onto or over the platform guardrail for any reason. Typically, small stepladders are provided to access the wing from the elevated platform. However, the lad-

der may not be climbed until the platform is raised so that the guardrail is just under the wing, thereby preventing a fall over the rail (Figure 8-19). Hoses, electrical cords and other obstacles must be kept clear from under the wheels of the lift.

Lighting: Lighting equipment used inside of fuel cells must be rated for use in a flammable/combustible environment and grounded. A number of different types of lighting devices are available, including 110- and 220-volt portable lamps, fiber-optic lighting systems, and explosion-proof flashlights or headlights (Figure 8-20). Lighting devices that provide adequate illumination for the job, but do not generate excessive heat, should be selected.

Other electrical equipment: As with lighting, any electrical equipment used inside fuel cells must be rated for a flammable/combustible environment and grounded. Electrical circuits should be ground fault circuit interrupter (GFCI) protected. In general, the use of electrical equipment inside fuel cells should be minimized.

Figure 8-19. Powered scissor lift under wing.

Photo courtesy of The Boeing Company

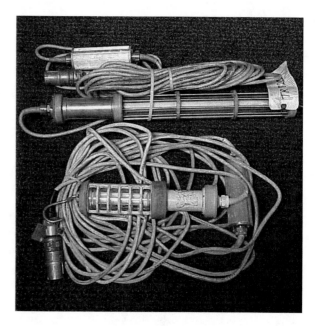

Figure 8-20. Explosion-proof portable lights.

Photo courtesy of The Boeing Company

Pneumatic equipment: Compressed air drive is preferred for power equipment used in fuel cells such as drills, grinders, rivet guns, seal guns and vacuum cleaners. It is essential that tools be connected to a source of clean air. Never use oxygen or inert gases such as nitrogen to drive air tools.

Hand tools: Hand tools taken into fuel cells should be made of non-sparking materials such as plastic, wood, aluminum or bronze. Tool containers should be plastic totes or cotton pouches.

Wipers, brushes, etc.: Activities such as mopping residual fuel, solvent cleaning, sealing and painting require the use of sponges, rags, gauze and brushes. Sponges are preferred for mopping out fuel. Rags used for solvent wiping should be made of cotton and kept as small as possible to help minimize the amount of chemicals used, and should be lint-free to avoid contamination of the fuel cell.

Waste containers: After use, wipers and brushes are contaminated with volatile chemicals. These items should be removed from the fuel cell or placed in a sealed container to prevent continuing evaporation that may contribute to a toxic or flammable atmosphere.

Fire extinguishers: A fire extinguisher must be ready to use in the immediate area. Some extinguishing agents may create a hazardous atmosphere inside of fuel cells. Consult with the local fire protection authority to select the size and extinguishing agent, and to specify procedures for safe use.

FOD control: Any item taken into a fuel cell has the potential to be left inside or to damage critical airplane components. Personal items and extraneous equipment not needed to accomplish the assigned task must be left outside. Required equipment should be inventoried before and after entry to ensure nothing has been left in the fuel cell.

Prohibited items: Specific items that never should be allowed inside fuel cells include cigarette-smoking materials, matches, lighters, food or drinks, chewing tobacco, metal buttons or belt buckles, loose-fitting jewelry, watches, pagers, mobile telephones and radios (except communication radios as required).

TRAINING

All personnel involved with fuel cell entry must receive specialized training. Courses must be designed specifically for aircraft fuel cell entry and must cover the activities conducted at a given facility. Some training content should be tailored to meet the needs of each role: entrant, attendant, detector operator, authorizer and emergency responder. However, a core of basic elements should be included for all. Note that the elements covered in this chapter are directed primarily toward ensuring personnel safety during entry. Additional technical information about working on the airplane structure and systems is required.

Basic fuel cell entry safety course content:

All designated participants in fuel cell entry operations should receive instruction in the following core areas:

- Definitions, including exposure terms, flammability terms, designation of fuel cells on subject airplane models, and definition of entry
- Hazards, including health, fire and physical hazards
- Signs of entrant distress or overexposure
- Entrant self rescue
- Fuel cell entry roles and responsibilities
- Communication systems
- Atmospheric monitor operation
- Ventilation techniques
- Personal protective equipment
- Tool and equipment requirements
- Emergency provisions
- Permit and logging requirements

Job simulation for entrants: After receiving core classroom instruction, fuel cell entrant trainees should be given the opportunity to practice an actual entry. An actual fuel cell (as dry as possible to minimize entry hazards) or a simulator may be used. Simulators may be constructed from plywood or other suitable material to the dimensions of a representative fuel cell (Figure 8-21). An advantage to the use of specially constructed training simulators is that they may be equipped with large escape doors or lids that allow easy egress if a trainee should have difficulty inside. Trainees should be guided through all permit-required steps, including setup of ventilation systems and air monitoring during the practice entry. Required personal protective equipment should be worn.

Training logistics: Instructors should be well-versed in both training techniques and fuel cell entry; instructors with actual entry experience are preferred. Trainees should be encouraged to ask questions throughout the process, and should be tested to ensure competency. Training documentation must include course title, and names of trainees and instructor. Initial training is required prior to assignment, and periodic refresher training in permit requirements is desirable. Additional training must be given as needed to cover changing processes or requirements.

Other required training: Fuel cell entry program participants usually require other safety training. The following list includes courses that may be required; check local safety and health regulations for applicability:

- Confined space entry
- Chemical hazard communication
- Respiratory protection
- Control of hazardous energy sources (lockout/tagout)
- Hearing conservation
- Use of personal protective equipment
- Fall protection

EMERGENCY RESPONSE

Employers of fuel cell entrants must have emergency response capability in place prior to any entry. Emergency responders may be employees or on-call from an outside agency such as the local fire department, or a combination of the two. Regardless of where emergency services will be obtained, advance planning is essential to ensure an adequate response.

Planning for emergencies: Emergency response must be part of the entrant employer's written fuel cell entry procedures. If emergency services will be obtained from an outside agency, the plan should include agreement regarding roles and responsibilities for each party. Up-front communication can help to avoid confusion during an actual emergency.

Qualification of responders: Fuel cell entry emergency response personnel must be familiar with the special hazards involved with fuel cell entry. Responders should receive training in the core elements described in the "Training" section of this document. In addition, responders should be trained in rescue technique and first aid/CPR. Responders must be physically capable of making entry and conducting rescue functions, including the removal of an incapacitated entrant. They also must be situated nearby and on-call so emergency response can be initiated in a timely manner. Fuel cell entry attendants may be designated as rescuers only if they receive additional specialized training, and then may enter to accomplish rescue only if relieved by another qualified attendant.

Figure 8-21. Fuel cell entry simulator.

Photo courtesy of Northwest Airlines

Emergency response equipment: Emergency responders must be equipped to enable safe entry under emergency conditions. The responder's basic inventory should include personal protective equipment, portable ventilation systems, and air monitors equivalent to those already specified in this document. In addition, the following equipment and supplies must be on hand:

• Self-contained breathing apparatus (SCBA) or air-supplying respirator systems are required if responders must enter fuel cells with "immediately dangerous to life and health" (IDLH) atmospheres. Under these conditions, an air-purifying respirator is not acceptable. Firefighters typically are equipped with SCBAs for entry into smoke-filled buildings; however the large size of the backpack mounted air tank makes them useless for fuel cell rescue. Remote tank SCBA systems are available and should be provided (Figure 8-22). Alternatively, air-supplying respirators connected to the plant compressed air system may be used, but only if an escape air supply is provided in case of air system failure. It should be noted that in nearly all emergency scenarios, auxiliary ventilation can quickly correct IDLH conditions in fuel cells.

• Tools as required to remove or break airplane components in order to assist recovery of a stricken entrant must be available.

• Life support/first aid equipment and supplies as appropriate must be available.

• Appropriate fire extinguishing equipment must be available. Note that some extinguishing agents have the potential to compromise the fuel cell atmosphere.

Summoning emergency services: A reliable system must be provided to allow the attendant to summon emergency services. Where telephones will be used, the emergency number should be shown prominently on the entry permit. Attendants should be trained in how to give clear and concise instructions regarding the nature and location of the emergency. Where telephones are not available, a substitute system such as two-way radios must be provided.

Figure 8-22. Remote tank SCBA.

Photo courtesy of The Boeing Company

Emergency procedures: During actual emergencies, absolute priority must be given to the safety of responders. Under no circumstances should any person be allowed to enter a fuel cell until all hazards are identified and controlled. All requirements specified in the original fuel cell entry permit continue to apply during the rescue, unless as-safe substitute procedures are developed and agreed to by the rescue team. Second priority is the rescue of any stricken entrant. It must be understood by all parties that aircraft or facilities may be damaged during rescue. Supervisors should stand by during an emergency to consult with the rescue team regarding aircraft structure and development of on-scene rescue procedures.

Rescue practice: Emergency responders should conduct practice rescue drills at least once per year. A rescue mannequin or volunteer should be recovered from a representative worst-case bay in an airplane or fuel cell simulator. All applicable procedures, including use of SCBAs or air-supplying respirator systems, should be followed. Other members of the fuel cell entry team should be allowed to observe each practice and participate in the post-drill critique.

ENTRY PERMIT AND PROCEDURES

Ensuring all potential hazards associated with aircraft fuel cell entry are addressed is a complicated endeavor. Rigorous control requirements must be met to protect entrant, airplane and hangar, and cooperation is essential between the various participants for each entry. This is why a written Fuel Cell Entry Permit and Checklist must be completed for each entry or series of similar entries. A well prepared comprehensive permit serves as a checklist to ensure that all requirements are met, designates the various roles by name, and provides documentation of what occurred once the entry is complete.

This section will discuss the line-by-line elements of the sample fuel cell entry permit that is included at the end of this chapter. The sample entry permit is a comprehensive example that can be used as a guideline in whole or part to develop a customized fuel cell entry permit.

Date issued: Fill in the date that the permit is prepared. A new permit should be prepared before each entry, or at the beginning of each shift for continuous operations. If stable operations are conducted repeatedly, a standing permit or standard operating procedure may be developed to cover multiple entries over an extended period of time. If a longer-term permit is used, entry logs, monitoring data sheets and a simplified checklist should still be prepared per entry.

Location information
1. Facility location of airplane: State the name of the airport or repair station; and use a hangar, bay or stall number to designate the exact location.

2. Situation of the airplane: Airplanes may be parked inside of hangars or on the ramp during fuel cell entry, as long as a location that will facilitate compliance with all entry requirements is chosen. Separation rules for open fuel cells, as stated in the airplane maintenance manual, must be followed. The location should be near emergency services, and uncluttered to allow free access for emergency vehicles and other equipment.

3. Securing the area: The airplane parking location must be demarcated as a controlled area in order to deny entry to unauthorized personnel and ensure that incompatible operations such as hot work or spray finishing are not conducted on or near the airplane. Barricades or barricade tape and signs should be erected around the perimeter of the controlled area.

Airplane information
4. Airplane model and tail number or other identification: Also specify airline if different airlines use this facility.

5. Boost pumps: Boost pump switches must be turned off and circuit breakers pulled and tagged to prevent damage and potential ignition sources.

6. Powered components: Powered components near the fuel cell entry point, such as 727 inboard leading edges, must be locked out and tagged.

7. Energy emitting systems: Communication and radar systems must be turned off on the entered and adjacent airplanes. Consult the airplane maintenance manual for separation distances between energized equipment and airplanes with open fuel cells.

8. Airplane electrical power: Ensure that there is no electrical power to the airplane during initial fuel cell entry. Power may be subsequently restored according to manufacturer's and operator's procedures.

9. Grounding and bonding: In order to eliminate the potential for static electrical discharges, airplanes must be grounded prior to opening fuel cells.

10. Location of fuel cells to be entered: Designate by bay (e.g., stub bay 3) or wing station (e.g., RH WS 70.5-114).

11. Defueling: Airplanes must be defueled using normal procedures. Tanks to be entered must be sumped by allowing all remaining fuel to flow from low point drains.

Process information

12. Purpose of entry: Describe the general activity (e.g., find and repair fuel leak).

13. Processes to be conducted: List specific processes (e.g., solvent clean, seal and topcoat).

Material information

14-18. Permitted chemicals: List all of the specific chemicals to be used inside such as solvents (e.g., MEK or MPK), sealants (e.g., ProSeal 890 or BMS 5-26), coatings (e.g., BMS 10-20), or any other material. Jet fuel is printed on the permit because some residual is always present.

Permitted containers: Specify the appropriate container for each chemical (e.g., plastic squeeze bottles for solvents and plastic tubes for sealants). Containers for liquids should be spillproof, with small openings to retard evaporation (Figure 8-23). Avoid containers such as metal cans, which may damage interior surfaces.

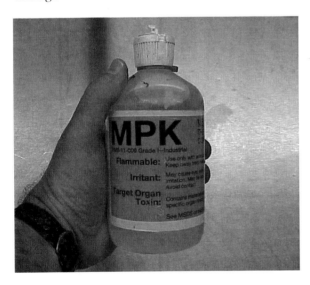

Figure 8-23. Solvent bottle.

Photo courtesy of The Boeing Company

Maximum use quantities: State the maximum quantity of each material allowed for each entry. Allowable quantities should always be limited to the smallest amount necessary to accomplish the task in order to minimize fire and exposure hazards. Work practices that minimize the potential for exposure to chemicals should be employed. When cleaning, solvents should be applied to the wiper, not squirted directly onto surfaces. Spray application of solvents and coatings should be avoided inside of fuel cells. As available, air monitoring data from prior entries which correlates volume used with resultant air concentration may be used to help set quantity limits.

19. Waste disposal: Used wipers and brushes will continue to evolve solvent vapors due to evaporation, which can contribute to a toxic or flammable atmosphere. These materials should be removed from the fuel cell as soon as they are generated, or placed in a vapor-tight container.

Hazard information

20-24. Hazard identification: Use the hazard information section to document the potential hazards identified during the hazard analysis. A typical entry would read, "Potential flammable atmosphere due to residual jet fuel and solvent use." Every hazard listed must be addressed by appropriate controls.

Ventilation requirements

25-26. Ventilation equipment: Specify the equipment to be used (e.g., "1,000 cfm Rhine Air venturi" or "50 m3/min Coppus blower and 10-inch duct").

27. Ventilation system bonding/grounding: All ventilation systems must be bonded to the airplane. Electrical fan motors must be grounded.

28-29. Pre-entry ventilation: Describe how the ventilation system will be configured for pre-entry ventilation or purging (e.g., exhaust venturis in RH top fill port and stub rear spar). Specify the length of time that the system must run prior to entry.

30-31. During entry configuration: Again, describe the configuration (e.g., blower duct in occupied stub bay, or adjacent bay for wings). Diagrams like those shown in Figure 8-10 may be helpful for more complicated set-ups.

Personal protective equipment

32. Respirator: Specify the respirator type, including type of facepiece (e.g., half-mask or full facepiece) and supplied air or cartridge type (e.g., organic vapor). If a sorbent cartridge respirator is used, specify the maximum use time.

33. Gloves: State the type of glove required (e.g. nitrile or flat film laminate) and the maximum use time.

34. Coveralls, cap and boot socks: Specify type and material

35. Eye and hearing protection: Specify type.

36. PPE verification: Condition of PPE should be checked before entry.

37. Breathing air quality: Breathing air must meet American National Standards Institute (ANSI) Grade D quality or equivalent specifications. This requirement usually can be met by using breathing air filter systems with carbon monoxide detectors (Figure 8-24).

Figure 8-24. Breathing air panel.

Photo courtesy of The Boeing Company

38-40. Atmospheric monitoring requirements: Air monitoring must be conducted prior to and during any entry. Use the accompanying Fuel Cell Air Monitoring Data Sheet to record results.

Attendant/communication/logging requirements

41. Attendant: Attendants must be provided for all entries. Attendant names must be recorded on the accompanying Fuel Cell Entry Log Sheet.

42-43. Communication: Entrant and attendant must be able to communicate at all times. Specify the communication system to be used (e.g., voice, tapping on structure, radio) and the agreed-upon interval for checking in (e.g., every 5 minutes). Verify function of electronic systems prior to entry. For non-verbal systems (e.g., tapping), entrant and attendant must agree on code prior to entry.

44. Logging: Each entry must be recorded on the Fuel Cell Entry Log Sheet, including entrant and attendant names, time in, time out, and location entered.

Other equipment requirements

45-46. Electrical equipment: All equipment used inside of or near open fuel cells must be explosion-proof, grounded and ground fault circuit interrupter (GFCI) protected.

47. Equipment static grounding: Metal entry stands and other fixed equipment should be grounded and bonded to the airplane to prevent buildup of a static potential.

48. Non-sparking tools: Hand tools used in fuel cells should be made of non-sparking materials such as brass, aluminum, wood or plastic.

49. Residual fuel: Fuel cells must be defueled and sumped prior to entry. Upon entry, workers should mop out any residual fuel left in the occupied area to prevent skin contact and continued evaporation of fuel vapor. It should be noted that some water-soluble fuel additives (especially biocides and icing inhibitors) may reach significant concentrations in the residual water left in fuel cells after draining. Avoid skin contact with residual water.

50. Inert gases: Inert gases in fuel cells can cause asphyxiation of the entrant. Verify that the supply line for pneumatic tools comes from a compressed air source.

Required training
51-54: Subject and number: List each required course by name and course number. "Fuel Cell Entry Safety" and "Respirator" are printed on the permit because they are required for any entry. Use the provided space to write in any additional required courses.

Required for: Use this column to designate which members of the entry team require the indicated courses.

Emergency provisions
55-56. Emergency services provider: Specify the in-house or outside provider and contact number.

57-58. Fire extinguishers: Specify type and location within 50 ft (15 m) of the airplane.

Other requirements
59-60. Space is provided for stating any additional entry-specific requirements.

Remarks/alternative requirements
Some entry jobs may require deviation from typical pre-entry procedures. In this situation, alternative procedures that provide equivalent protection for the entrant must be developed. Document the alternative procedure for any checklist item marked "No" by line number. This space also may be used for additional remarks.

Verification
61-63. Verification and signature: The entry authorizer must verify that the permit and checklist is complete. By signing, the authorizer attests that all permit requirements have been met and authorizes entry.

Fuel Cell Air Monitoring Data Sheet
The Fuel Cell Air Monitoring Data Sheet accompanies the permit. It is used to systematically record the data gathered from atmospheric monitoring. A custom sheet must be developed if readings will be taken for contaminants other than jet fuel and oxygen.

General information
Atmospheric monitor: Record the make and model of the detector used. Also record the equipment number; manufacturer's serial number or owner-assigned unique equipment numbers may be used.

Calibration data: Record the date of the last calibration, and the calibration gas or gases used.

Conversion factors: Most detectors require the use of a conversion factor when measuring jet fuel vapor. Find the appropriate conversion factor in the detector manufacturer's manual and record it on the data sheet. Note that the conversion factors for percent LEL and ppm readings usually are different.

Detector operator's name: Record on the data sheet. The detector operator usually is the attendant or the entrant. If the entrant is monitoring the space, readings should be called out to and recorded by the attendant.

Air Monitoring Data
Time and location: Record for every reading. The top two lines of the data sheet are reserved for ambient and pre-entry readings. Typical location entries would read "inside stub access door" or "WS 114 at front spar".

Percent oxygen readings: Record in the "%O_2" column. Investigate any significant variance (+/-0.5 percent) from the ambient concentration (usually 20.9 percent). Evacuate the fuel cell if the oxygen concentration is less than 19.5 percent or more than 23.5 percent.

Percent of jet fuel vapor lower explosive limit: Use the "% LEL Jet indicated" column to record LEL readings as shown on the instrument display. If required, multiply each reading by the appropriate correction factor and record result in the "% LEL Jet corrected" column. Evacuate the space if true % LEL readings exceed 10 percent.

Parts per million of jet fuel vapor: Use these columns to record ppm readings for jet fuel vapor. As with percent LEL, a conversion may be required. Evacuate the space if the applicable exposure limit is exceeded and no respirators are in use. If respirators are worn, evacuate the space if the product of exposure limit times respirator protection factor is exceeded.

Fuel Cell Entry Log Sheet

Use this sheet to record data for each individual entry. The date, times of entry and exit, entrant's and attendant's names, and entry location must be recorded.

Posting and retention of permits: The active permit should be posted in a conspicuous spot in the work area (Figure 8-25), or held by the attendant. Once the entry is complete, or if entry is suspended because conditions varied outside of those permitted, the permit and attachments should be closed out and retained for at least one year for program assessment purposes. Permits and attachments that contain employee chemical exposure data should be retained longer.

SPECIAL OPERATIONS

This section covers certain special operations occasionally conducted inside of fuel cells. These operations are covered separately because they may compromise one or more of the basic safety requirements for fuel cell entry. Special alternative procedures are needed to ensure an equivalent level of safety. In general, these and other special fuel cell operations should be conducted only as allowed by the airplane manufacturer and internal procedures.

Negative pressure leak detection: This process pinpoints the location of hard to find fuel leaks by applying a partial vacuum to the fuel cell while an

Figure 8-25. Entry permit placard.

Photo courtesy of The Boeing Company

entrant is inside. The entrant uses soap bubble solution to locate the leak. Because the fuel cell must be sealed to apply vacuum it cannot be ventilated during the process. The following special requirements apply:

- The fuel cell must be mopped completely dry, and thorough pre-entry ventilation must be completed.
- Atmospheric readings for percent oxygen and percent LEL (after application of appropriate conversion factor) must match ambient air after ventilation is removed. Ppm readings should be minimal.
- The temporary closure for the access hole should be transparent. It must be removable from inside and outside, and a manual vacuum relief valve should be operable from inside and outside. The closure must also be equipped with a vacuum gauge and an automatic vacuum relief valve set for 2.5 psi or less. Fittings for breathing air connections also are required. Connections for atmospheric monitors and/or hard-wired communication systems may be added if required.
- The entrant shall wear a supplied air respirator.
- Continuous atmospheric monitoring shall be conducted inside the fuel cell throughout the negative pressure entry. Verify with the manufacturer that the monitor will function correctly in a negative

pressure situation, and adjust oxygen alarm set-point as appropriate for reduced pressure.

- Entrant and attendant shall be equipped with an electronic communication system.
- Soap solution and marker pens are the only chemical products that may be taken into the fuel cell. No other materials or processes are allowed during a negative pressure leak check.
- Pressure differential shall be limited to less than 2 psi. Vacuum shall be applied and relieved slowly to avoid trauma to the entrant's ears.

Hot work: Elimination of potential ignition sources is essential to controlling fire and explosion hazards during fuel cell entry. However, some processes such as structural rework or fastener removal may involve spark-producing operations. These operations rely on other procedures to eliminate the potential for fire:

- A hot work permit must be issued by a qualified person familiar with aircraft fuel cell fire protection issues. This permit ensures that special controls are deemed adequate by a competent expert.
- The fuel cell must be mopped completely dry, and thorough pre-entry ventilation must be completed.
- Identify and close any open channels to vapor-containing spaces, such as fuel plumbing or tank vents.
- Continuous atmospheric monitoring shall be conducted inside of the fuel cell throughout the hot work process.
- Atmospheric readings for percent oxygen and percent LEL must match ambient air. Ppm fuel vapor readings should be minimal.
- During-entry ventilation must be sufficient to ensure that zero percent LEL readings are maintained.
- No flammable chemicals or combustible materials may be taken into the fuel cell during hot work.
- Atmospheric hazards resulting from burning sealant and paints must be considered.
- A fire watch shall be maintained during all fuel cell hot work.

Spray application: Spray application of fuel cell coatings greatly intensifies atmospheric hazards. Spraying causes the solvents in the coating to evaporate much faster, and also creates a mist that may contain chromate compounds – which are known human lung carcinogens – or other hazardous air contaminants. Brush application of coatings is always preferred. If spraying must be conducted:

- Continuous atmospheric monitoring shall be conducted inside of the fuel cell throughout the spraying process.
- During-entry ventilation must be sufficient to ensure safe levels are maintained.
- Personal protective equipment capable of protecting the entrant from all components of the coating must be selected.
- High volume low pressure (HVLP) spray equipment, adjusted to minimize overspray, shall be used.

Cryogenic applications: Certain fasteners or "freeze plugs" must be chilled to very low temperatures to achieve critical dimensions just prior to installation. Liquid nitrogen or dry ice typically are used for this purpose. When these cryogenic materials are exposed to normal atmospheric conditions, they quickly revert to their normal gaseous state. If this occurs inside of a fuel cell, the resultant inert gas infusion may dilute oxygen to the extent that asphyxiation is possible. Freezing with liquid nitrogen or dry ice should be accomplished outside of the fuel cell whenever possible. If these materials must be taken inside, these procedures apply:

- Minimize the amount of cryogenic material taken into the fuel cell to the least amount necessary.
- Use only spill-proof containers for liquid nitrogen.
- Continuous atmospheric monitoring shall be conducted inside of the fuel cell throughout the freezing process.
- During-entry ventilation must be sufficient to ensure safe levels of oxygen are maintained.
- The entrant shall wear a supplied-air respirator.

Fuel system functional test: Occasionally a technician may need to check fuel system function or look for leaks in pressurized fuel plumbing while inside a fuel cell. This practice introduces additional hazards because it may require that in-tank electrical components such as boost pumps be energized, and because leaks may generate a mist of fuel within the space. Every effort should be made to verify correct system function from outside of the fuel cell. If entry with active systems must be made:

- Consult the maintenance manual and engineering to develop a test plan that will ensure the safety of the crew and airplane.
- Effective, constant communication between flight deck system controller, attendant and entrant must be maintained.
- A full facepiece supplied air respirator shall be worn.
- Personal protective equipment including fuel resist-ant coveralls, neoprene booties and Nitrile gloves shall be worn.
- Continuous atmospheric monitoring shall be con-ducted inside of the fuel cell throughout the test.
- During-entry ventilation must be sufficient to ensure that safe levels are maintained.
- De-pressurize the fuel system as soon as function is verified or the leak is located.
- If an uncontrolled leak occurs, the entrant shall exit the fuel cell immediately.

REFERENCES

U.S. Occupational Safety and Health Administration (OSHA). *1910 Subpart J, Confined Spaces*

American Conference of Governmental Industrial Hygien-ists. *Threshold Limit Values for Chemical Substances*

American National Standards Institute. *Z117.1-2003, Safety Requirements for Confined Spaces*

WORKING GROUP MEMBERS

Chairman:

Bryan Bauer - Boeing Commerical Airplane Group

Members:

Wilton Arizmendi – United Airlines

Ron Failing – Frontier Airlines

Norman Hogwood – AeroGround Safety Services Inc

Tom Jassoy

Mike Krause – Clayton Group

Chuck Lovinski – F. Rine & Assoc

Michael Lueck – ABXAir Inc

Roland (Jim) Page – Page Safety Solutions

Tom Pazell – United States Air Force

Everson I. Pereira – Singapore Airlines

Les Rose – Saudi Arramco

David Sorrels – Boeing Commercial Airplane

Terri Weiland – TR Weiland Company, LLC

AIRCRAFT FUEL CELL ENTRY PERMIT AND CHECKLIST

	Instructions: Fill out all WHITE areas on both sides of this form. Also complete the accompanying FUEL CELL ENTRY LOG SHEET and AIR MONITORING DATA SHEET.		Date:		

#	LOCATION INFORMATION			YES	NO*
1	Facility Location of Airplane:				
2	Is the airplane positioned to allow uninterrupted entry and free movement of equipment?				
3	Is the entry area secured, with OPEN FUEL CELL signs posted?				

#	AIRPLANE INFORMATION			YES	NO*
4	Airplane Model:	Airplane Tail No. or Other Identification:			
5	Are boost pump switches OFF, with circuit breakers pulled and placarded?				
6	Are moveable powered components near the entry point (e.g. B-727 inboard leading edge) locked out?				
7	Is radar and communication equipment on this airplane and nearby airplanes turned OFF?				
8	Are batteries and external power disconnected and placarded?				
9	Is the airplane grounded?				
10	Location of Fuel Cells to be Entered:				
11	Are tanks to be entered de-fueled and sumped?				

#	PROCESS INFORMATION			YES	NO*
12	Purpose of entry:				
13	Processes to be Conducted:				

#	MATERIAL INFORMATION: Only the materials listed here may be used.			YES	NO*
	Permitted Chemicals	Permitted Containers	Maximum Use Quantity		
14	Jet Fuel A	N/A	Residual only		
15					
16					
17					
18	Are the permitted chemicals on hand and in the permitted containers? (mark each line above)				
19	Are conductive plastic bags or other sealable containers provided for used wipers?				

#	HAZARD INFORMATION (List)			YES	NO*
20					
21					
22					
23					
24	Have all potential hazards been identified and addressed?				

#	VENTILATION REQUIREMENTS			YES	NO*
25	Ventilation Equipment:				
26	Is the required explosion proof ventilation equipment on hand?				
27	Is all ventilation equipment bonded to the airplane?				
28	Pre-entry Configuration:	Pre-Entry Vent Time:			
29	Has the pre-entry ventilation system run for the prescribed time?				
30	During-Entry Configuration:				
31	Is the during-entry ventilation system properly configured and running?				

**Caution: Fuel cells must be ventilated the entire time that they are open.
All work shall stop and the fuel cell shall be evacuated it the ventilation system fails,
or if any ill effects such as dizziness, irritation, or strong odors are noted by the entrant.**

#	PERSONAL PROTECTIVE EQUIPMENT REQUIREMENTS		YES	NO*
32	Respirator:	Max. Use Time:		
33	Gloves:	Max. Use Time:		
34	Coveralls, Cap, Boot Sox:			
35	Eye Protection:	Hearing Protection:		
36	Is the required personal protective equipment on hand and in good condition? (mark each line above)			
37	Has respirator breathing air quality (if required) been verified?			
#	**ATMOSPHERIC MONITORING REQUIREMENTS**		**YES**	**NO***
38	Is there a calibrated atmospheric monitor on hand and ready for use?			
39	Is the FUEL CELL AIR MONITORING DATA SHEET attached with the top section filled out?			
40	Have acceptable pre-entry readings been taken and recorded?			
#	**ATTENDANT/COMMUNICATION/LOGGING REQUIREMENTS**		**YES**	**NO***
41	Has an attendant been provided to look after the entrant?			
42	Communication System:			
43	Has the communication system been tested, and is it understood by entrant and attendant?			
44	Is the FUEL CELL ENTRY LOG attached and ready for use?			
#	**OTHER EQUIPMENT AND MATERIAL REQUIREMENTS**		**YES**	**NO***
45	Is all lighting and other electrical equipment to be used in or near fuel cells rated explosion proof?			
46	Is all electrical equipment grounded and GFIC protected?			
47	Are entry stands and other fixtures grounded and bonded to the airplane?			
48	Are hand tools made of non-sparking materials?			
49	Are sponges or rags provided for the entrant to mop out the fuel cell work area?			
50	Has the supply system for pneumatic tools been verified as providing air (no N2 or other inert gas)?			

#	TRAINING REQUIREMENTS			YES	NO*
	Subject	Course No.	Required for:		
51	Fuel Cell Entry Safety		Entrant, Attendant, Det. Operator, Supervisor		
52	Respirator Use		Entrant		
53	Other (specify):				
54	Have entry team members received required training? (mark each line above)				

#	EMERGENCY PROVISIONS **Caution: Only trained fuel cell emergency personnel may enter fuel cells to perform rescue.**		YES	NO*
55	Emergency Services Provider:	Phone Number:		
56	Are emergency services available if needed?			
57	Fire Extinguishers; Type and Location:			
58	Are the correct fire extinguishers available?			
#	**OTHER REQUIREMENTS (SPECIFY)**		**YES**	**NO***
59				
60				

#	REMARKS/ALTERNATIVE REQUIREMENTS *Document as-safe alternative requirements for any checklist items marked NO by line number.

#	VERIFICATION		YES	NO*
61	Are all of the WHITE areas of this permit filled out?			
62	Is there an as-safe alternative listed above for any checklist item marked NO ?			
63	Signature: I confirm that all of the requirements listed on this permit are met:			
	Authorizing Supervisor or Designee:	Printed Name:	Date:	

AIR MONITORING DATA SHEET

GENERAL INFORMATION

Atmospheric Monitor:			Serial No:	
Calibration Date:	Calibration Gas:	Conversion Factor (LEL):	Conversion Factor (ppm):	
Detector Operator´s Name:				

AIR MONITORING DATA

Time	Location	% O2 (1)	LEL Jet indicated	LEL Jet corrected (2)	ppm Jet indicated	ppm Jet corrected (3)
	Instrument Test: Outside fuel cell in ambient air					
	Pre-Entry:					

(1) **Percent oxygen concentration. Investigate any variance from ambient. Evacuate fuel cell if less than 19.5% or greater than 23.5%.**

(2) **Percent LEL of Jet A fuel vapor, corrected by multiplying indicated value by correction factor, if any. Evacuate fuel cell if greater than 10%.**

(3) **Parts per million of Jet A fuel vapor, corrected by multiplying indicated value by correction factor, if any. Require respiratory protection if exposure limit is exceeded. Evacuate fuel cell if exposure limit times respirator protection factor is exceeded.**

Use additional sheets for other contaminants as necessary.

FUEL CELL ENTRY LOG SHEET					
Date	Time In	Time Out	Entrant's Name	Attendant's Name	Tank Entered

FALL PROTECTION

CHAPTER 9

I n the aviation industry, there are numerous fall-from-height hazards. Servicing, maintaining and repairing commercial aircraft of all types requires that workers access parts and structures using ladders, lifts and other types of elevated platforms. In some cases, the worker will step onto the surface structure of an aircraft, such as the wing or horizontal stabilizer. In every case, it is necessary to recognize the fall hazard and develop appropriate fall protection measures.

This chapter will describe the elements of an aviation fall protection program, including hazard identification, risk assessment, protection method selection, worker training and rescue operations.

FALL PROTECTION PROGRAM ELEMENTS

An effective aviation fall protection program is best administered with a documented management system that specifies requirements and responsibilities. The written program should describe:

- Where and when fall protection is required
- Who is responsible for various program elements and what those responsibilities are
- How fall protection controls will be selected and administered
- Employee training and emergency response requirements

RISK MANAGEMENT

Risk management is defined as a continuous process of detecting, assessing and controlling risk. For purposes of providing an effective fall protection program, the following process can be applied.

Step 1. Identify the hazard.

Step 2. Assess the risk.

Step 3. Analyze risk control methods.

Step 4. Make control decisions.

Step 5. Implement risk controls.

Step 6. Supervise and review.

FALL HAZARD IDENTIFICATION

Falls can occur at any time and in any place. An effective program will detect when fall protection is necessary due to regulatory requirements, potentially high probability or severity, or a combination of these factors.

Regulatory requirements
In general, government regulations require that fall protection be provided in specific circumstances and at specific heights. These circumstances and heights may vary depending on the jurisdictional authority. Careful review of governing regulations must be conducted to ensure that the company is in compliance with relevant requirements.

Probability factors
Various factors may increase the probability of a fall occurring. These factors include slippery footing, uneven and elevated work surfaces, numerous objects to trip over, and physiological disorders. The presence of any of these factors signifies a need for further evaluation.

Figure 9-1. Tail fin work using a boom lift.

Photo courtesy of The Boeing Company

Figure 9-2. Wing guarding tool.

Photo courtesy of The Boeing Company

Figure 9-3. Overhead fall arrest for engine maintenance.

Photo courtesy of The Boeing Company

Severity factors

Determining the potential severity of a fall is difficult, because even a fall on the same level may result in a serious or even fatal injury. However, in general, the farther a person falls, the more potentially severe the injury will be. Activities taking place more than 4 ft above the next lower surface require further evaluation.

Specific aircraft operations

Numerous aircraft activities may create a fall hazard for personnel. These include:

- Cleaning and washing
- Heavy maintenance and overhaul
- Painting and stripping
- Deicing
- Engine maintenance

RISK ASSESSMENT

Assessing the risk of a fall involves quantifying the probability and severity factors previously identified. There are other factors that can also be assessed. These include:

- Proximity: How close to the edge of an elevated platform is the worker?
- Duration: How long is the worker exposed to the hazard?
- Frequency: How often does exposure occur?
- Number of workers: How many workers are exposed to the hazard?
- Environment: What kind of environmental conditions are present that may cause slips, trips or loss of balance?

Table 9-1 shows a sample risk assessment matrix accounting for the previously described fall hazard probability and severity factors. This matrix can be used to assign relative levels of risk for identified fall hazards, and to help prioritize additional fall protection activities.

PROTECTION METHODS

The process of selecting the appropriate fall protection method is similar to the approach needed to control other hazards. First, eliminate hazards that can

To assess the level of risk, the potential severity of the injury incident and the estimated probability of its occurrence must be determined.

For fall hazards, the severity of the potential injury is determined by the height of the fall.

The probability of the incident occurring is influenced by several observable conditions. These conditions can be quantified using the point scoring system described in the table.

		RISK LEVEL			
SEVERITY	**Catastrophic** Height is greater than 12 ft	EXTREME	EXTREME	HIGH	HIGH
	Critical Height is between 8 and 12 ft	HIGH	HIGH	MODERATE	MODERATE
	Moderate Height is between 4 and 8 ft	HIGH	MODERATE	MODERATE	LOW
	Negligible Height is below 4 ft	LOW	LOW	LOW	LOW
		Frequent More than 8 points	**Likely** 6 to 8 points	**Occasional** 3 to 5 points	**Seldom** 2 points or less

PROBABILITY

Proximity: How close to the edge is the worker?
More than 6 ft away? add 0 points
From 3 to 6 ft away? add 1 point
Within 3 ft? add 2 points

Duration: How long is the worker exposed to the hazard?
10 minutes or less? add 1 point
More than 10 minutes? add 2 points

Frequency: How often does the exposure occur?
Less than weekly? add 0 points
Weekly? add 1 point
Daily? add 2 points

Number of workers: How many workers are exposed to the hazard?
One worker only? add 1 point
More than one worker? add 2 points

Environmental conditions: Is the worker exposed to windy conditions or materials that create a slip or trip hazard?
Working outside? add 1 point, and
Slip or trip hazards present? add 1 additional point

MAXIMUM PROBABILITY SCORE = 10 POINTS

Table 9-1. Fall protection risk assessment matrix.

be eliminated. Next, minimize and control the hazard by providing engineering controls, such as guarding and barriers. Finally, provide the necessary personal protective equipment as the final choice in hazard control methods.

Housekeeping

Maintaining a clean and orderly work area is the first rule of fall prevention. Eliminating the hazards that can cause slips and trips is a worthwhile but never-ending job. Every organization should have a well-managed housekeeping plan that sets out daily requirements for maintaining a clean and orderly workplace.

Guarding

The preferred method of providing fall protection for elevated work platforms is with passive guarding systems. Standard guardrails provide substantial protection against a fall to a lower level. U.S. Federal OSHA regulation requires that guardrails be provided for all elevated work platforms more than 4 ft higher than the next lower level. Specifications for the construction of guardrails are contained in the regulations.

In aircraft operations, open-sided elevated work platforms are often used with the open side of the platform next to the aircraft, as seen in Figure 9-4. In many cases, these types of platforms will incorporate temporary barriers and floor covers to protect the openings in the walls and floor of the platform. These provide fall protection when the work platform is not positioned snugly against the side of an aircraft.

Vehicle mounted lift platforms also typically incorporate guardrail systems around the work platform of the lift. Articulating boom lifts may have either guardrails or fully enclosed baskets. Scissors lifts typically use guardrails. On some lifts, the guardrails may be collapsed or removed for specific activities.

Door barriers are often used during maintenance activities when aircraft doors must remain open with no work stands or steps in place. Lower level maintenance compartments access points can also create a fall hazard if left open and not guarded.

Figure 9-4. Open sided work platform.

Photo courtesy of The Boeing Company

Personal restraint systems

Personal restraint systems are designed to keep a worker from being able to reach and fall over an unguarded edge. Restraint systems can utilize a body belt, chest style harness, or full body harness. However, different jurisdictions have disallowed the use of body belts and chest style harnesses for fall protection, and require that only full body harnesses be used in personal fall protection systems.

The worker connects to the restraint system anchor point using a lanyard of predetermined length. The length of the lanyard in a restraint system should be sized so as to prevent the worker from reaching an unguarded edge.

The anchor point in a restraint system should be capable of supporting the maximum expected load created by the system, with an appropriate safety factor. Although there are no specific regulations that define these requirements, it is generally accepted that a 4 to 1 safety factor for restraint systems is appropriate, and that the maximum expected load in such a system will typically not exceed 200 pounds (91 kg). Using these criteria, the anchor point in a personal restraint system should be capable of supporting at least 800 pounds (364 kg).

Fall arrest systems

Personal fall arrest systems are similar to restraint systems in that they use similar equipment to connect the worker to an anchor point. A fall arrest system, however, is designed to keep the worker from striking a lower surface after a fall occurs, and to absorb some of the impact forces that are created during the fall event.

A fall arrest system must utilize a full body harness. Using a body belt or chest style harness is not permitted by government regulation, and may subject the user to serious internal injuries due to the inadequate distribution of arrest forces across the body. A full body harness is designed to transmit the jarring arrest forces to the stronger pelvic part of the body, and more evenly distributes these forces upon impact.

A fall arrest system must also incorporate shock-absorbing capabilities. The arrest forces created during a fall event are dependent upon the weight of the individual and the height of the fall before the fall arrest sequence is initiated. A 200-pound worker using a system without shock absorbing capability and falling 6 ft can generate in excess of 2,200 pounds of force. Without shock absorbing capabilities in the fall arrest system, the body must absorb the force. A high level of arrest force easily can cause serious injury even if the fall is arrested and the worker does not strike a lower surface.

Therefore, most government regulations have established a maximum arresting force that can be transmitted into a worker's body by a fall arrest system. In the United States, this is 1,800 pounds. To meet this maximum arrest force requirement, the lanyard used to connect from the harness to the anchor point must include a built-in or added-on shock-absorbing component, or must be constructed so that the arrest forces do not exceed 1,800 pounds.

In most cases, fall protection equipment manufacturers can provide lanyards and other devices that will meet these requirements. Indeed, many manufacturers offer shock-absorbing equipment that limits the arrest forces to no more than 900 pounds, reducing by half the amount of force that can be transmitted into the worker's body.

The final part of a personal fall arrest system is the anchor point. For a fall arrest system, the anchor point must be capable of supporting either 5,000 pounds or, for an engineered system, two times the maximum expected load. For example, the anchor point in an engineered

system using shock absorbing components that limit the maximum arresting force to no more than 900 pounds must be capable of supporting at least 1,800 pounds.

Self-retracting lifelines

A self-retracting lifeline is a special purpose device with many applications in the aviation industry. An example of a self-retracting life is shown in Figure 9-5.

These devices provide an extendable line kept under constant tension by a spring in the lifeline housing. If a worker falls, the device is designed to nearly instantaneously stop deployment of the line, arresting the fall within just a few inches.

Most self-retracting lifelines also include a shock-absorbing feature, usually a clutching mechanism inside the housing. These characteristics significantly reduce the arrest forces impacting the fallen worker.

The other advantage of using a self-retracting lifeline in an overhead system is the capability of the system to allow up and down movement of the worker through a wide range of height. The self-retracting lifeline device can also be mounted on a beam and trolley system, which allows workers to move back and forth, for example, on top of a wing. For this reason, a fall arrest system

Figure 9-5. Self-retracting lifelines.

Photo courtesy of The Boeing Company

using self-retracting lifelines is the preferred fall protection method for suspended work platforms and over-wing installations in aircraft hangar facilities.

Vacuum pad anchor devices

Vacuum pads are designed to be able to attach to a smooth aircraft surface structure, providing a temporary anchor point for either a personal restraint or fall arrest system. Examples of vacuum pad anchor devices are shown in figures 9-6 and 9-7.

The vacuum pad works by sealing against an appropriate surface. The pad is then stabilized by evacuation of the atmosphere under the pad. This creates a negative pressure, holding the pad in place. Currently available vacuum pads have weight capacities ranging from 1,500 pounds up to nearly 4,000 pounds.

The vacuum pad anchor device is used in conjunction with standard harnesses and connecting lanyards. For vacuum pad anchors used in a restraint system, it is critical that a properly sized lanyard be used. For vacuum pad anchors used in a fall arrest system, the connecting lanyard must include a shock-absorbing component. Use of a vacuum pad anchor device requires thorough training, as the characteristics of the devices must be understood completely.

The advantage of a vacuum pad anchor system is the flexibility that it allows. The vacuum pad can be installed on nearly any smooth surface on the aircraft, as long as the structure of the aircraft can support the expected forces transmitted back into the anchor

Figure 9-7. Vacuum pad anchor system.

Photo courtesy of The Boeing Company

device. The anchor devices also can be quickly and easily installed and removed, making it convenient to provide protection for short duration jobs.

TRAINING

Worker training must be conducted for all types of fall protection systems. Workers should be trained in how to install and remove temporary guardrail systems and barriers. Workers using vehicles or lifting equipment must be trained in how to safely operate and use this equipment. Finally, workers using any type of personal fall protection equipment must be trained in the safe and correct way to use this equipment. Failure to train workers properly is a significant factor in many fall related incidents.

Personal protection systems

Training on personal fall restraint and fall arrest systems must include detailed information on the following:

- How to inspect and care for various system components, including harnesses, lanyards and other connecting equipment
- How to put on and fit a full body harness
- How to install and use anchor point equipment
- The limitations and restrictions of the various systems
- Self-recovery techniques when a fall occurs
- Emergency response procedures

Figure 9-6. Vacuum pad anchor device.

Photo courtesy of The Boeing Company

Vehicle and equipment operation

Employees using articulating and boom lifts, scissors lifts, and other types of elevating platforms must be trained in the safe use and operation of this equipment. In addition to safe operations, this training should include:

- Limits of the equipment with respect to tipping over and stability
- Specific fall protection features of each piece of equipment
- How to return safely to the ground if the equipment malfunctions

RESCUE AND EMERGENCY RESPONSE

As with any industrial activity, emergency situations can and do occur in aviation ground operations. In dealing with falls, unique conditions exist and – without adequate planning – companies may be unprepared to deal with fall emergencies.

Planning

The first step in fall incident rescue and emergency response is planning. A properly developed fall incident emergency response plan will include information on:

- Notification: How are responders summoned to the scene?
- Responder qualifications: What unique abilities and skills do responders to fall incidents need?
- Rescue procedures: Are there special procedures needed for rescuing workers who may be suspended or who may be incapacitated at an elevated position?
- Response equipment: Is there special equipment or devices that may be needed to perform rescue?

These questions must be answered before a fall emergency incident happens. Without prior planning, a minor incident may develop into a serious or even catastrophic event.

Rescuing suspended workers

The most critical element of a fall incident emergency response plan is to make sure that suspended work-

ers can be rescued quickly. A worker attached to a fall arrest system may not be able to self-recover back to a stable and upright position, or may be injured by the impact of the fall arrest forces and be incapacitated. A fallen worker suspended in a full body harness for a prolonged period can suffer injury due to the constriction of blood flow and breathing capacity. The ability of a person to remain suspended without suffering injury is dependent on body weight, physical condition and other factors, but in no case should a worker be left in a suspended position for more than 15 minutes. After 15 minutes, the risk of suffering further injury is greatly increased. Devices are available to facilitate self rescue.

REFERENCES

U.S. Occupational Safety and Health Administration (OSHA):
  1910 Subpart D, Walking-Working Surfaces.
  1910 Subpart F, Powered Platforms, Manlifts, and Vehicle-Mounted Work Platforms.
  1926 Subpart M, Fall Protection Systems Criteria and Practices.
American National Standards Institute:
  Z359.1 – Personal Fall Arrest Systems, Sub-systems and Components.
  MH29.1 – Industrial Scissors Lifts.
  A92.2 – Vehicle Mounted Aerial Devices.
  A92.5 – Boom Supported Elevating Platforms.
  A92.7 – Airline Ground Support Lift Vehicles.

WORKING GROUP MEMBERS

Chairman:

David Sorrels - Boeing Commercial Airplane

Members:

John Kane – Air Canada

Laura Quigley – Delta Airlines

Kevin Frommelt – Northwest Airlines

Bill Simpson – Piedmont Hawthorne

Bill Carlyon – Boeing Commercial Aviation

GROUND SUPPORT EQUIPMENT

CHAPTER 10

Equipment design considerations
Controls
Brakes
Stability
Fuel system
Operator's compartment
Hoisting/lifting
Vehicle working surfaces
Maneuverability
Exhaust systems
General
Equipment maintenance and operation
Maintenance
Operation
Personnel considerations
Personnel qualifications
Personnel responsibilities
Training programs
Signaling
Catering operations
Food service handling
Operational guidelines
Qualifications
General airport driving rules
Loading
Positioning trucks at aircraft
Raising truck bodies
Servicing an aircraft
Leaving the aircraft
Special procedures and precautions
References

Ground support equipment (GSE) includes both motorized and non-motorized units that are needed to service aircraft safely and efficiently. This chapter lists design and procedural considerations in procuring, operating, and maintaining mobile ground equipment used in airports.

In the selection of proper ground support equipment, the function that the equipment is to perform must be evaluated. In evaluating the function, consider these factors:

- Types of aircraft to be serviced
- Possible use other than aircraft servicing
- Use of the equipment at other airports or on public highways
- Minor changes or modifications that would allow for use in other ways than originally intended.

These considerations may reduce or eliminate the need for additional pieces of ground support equipment. All changes or modifications should be subjected to an approval process that may include the manufacturer or others.

EQUIPMENT DESIGN CONSIDERATIONS

All equipment should conform to appropriate standards, such as those of the European Standardization Committee (CEN) or the American National Standards Institute (ANSI), the International Organization for Standardization (ISO), and the International Electrotechnical Commission (IEC). (See References at the end of this chapter.)

Controls

Controls and controlling circuits should be designed in such a manner that failure within a control or its circuitry will not introduce an unsafe operating condition.

Controls should be grouped and located so that the operator can easily reach them with his or her operational hand.

The design and location of such controls should be in a manner that does not permit clothing to catch on them.

Controls should be sized and spaced to provide easy operation with a gloved hand.

Controls should be identified with permanently affixed and nonfading placards using internationally recognized symbols. The lettering or symbols on the placards should be easy to read from the operator's position and should contrast sharply with the background. Signs should indicate the function or motion of the control.

Controls and instruments exposed to the weather should be weatherproof. They should be protected from snow and ice accumulation.

Operations control and instrument panels, exclusive

of driving controls, should be illuminated and not produce a glare to the operator.

Ground support equipment with automatic transmission shall have the transmission shift sequence location conforming to relevant international standards.

Transmission shift should be sized to be handled with a gloved hand with unobstructed clearance in any position.

All other operational controls should move in the direction of travel for the function that they control. They should be indented or similarly locked into the operating positions to prevent inadvertent deactivation or reversing. Such controls should be readily available to the operator(s). Emergency-only operating controls need not meet this requirement.

On/Off switches should be "on" when in the up position, or if mounted horizontally, they should be "on" when they are away from the operator.

Clutch, foot brake, accelerator, and other functional foot controls should be provided with a slip-resistant material.

Foot controls should be spaced to provide easy operation with a booted foot.

Brakes

Based on the vehicle brake design, two systems should be provided for applying brakes:

1. The primary system (service brakes) should be pedal-operated to apply brakes simultaneously to all braked wheels.
2. The secondary system (parking brakes) should be hand-operated to apply a braking action sufficient to hold the vehicle on the maximum grade that the vehicle can negotiate with rated capacity load, with the transmission in neutral and the engine idling on a dry, clean concrete surface.

The brake should be effective in restraining vehicle movement to either a forward or a reverse direction.

Stability

All ground support equipment should be stable when exposed to wind and/or jet blast loads. Units not designed for such winds, such as rollup boarding stairs, should be removed from the ramp or secured when high winds are predicted.

Lift units exposed to jet blast, wind, and irregular surfaces should be equipped with stabilizing devices, if necessary, that would preclude their overturning when exposed to wind or blast.

Retraction of the stabilizing device should not be possible under normal or emergency conditions, until the unit has been lowered to within the stability previously discussed.

Stabilizer-activating devices should be located so as not to expose the operator to injury.

Emergency stabilizer-raising controls should be provided and should be located so as not to expose the operator to injury.

Stability devices that extend beyond the vehicle profile should be painted yellow or illuminated.

Fuel system

Fuel tanks should be located for maximum protection from collision damage.

A clearly visible permanent marking that indicates the type of fuel to be used should be installed adjacent to the filler.

Fuel tank fillers should be located outside the vehicle, vent directly to free atmosphere, and be accessible from ground level.

Operator's compartment

If a vehicle has an enclosed cab, the cab should provide an ergonomically friendly environment to include proper ventilation, temperature and noise reduction to recommended levels. For open cab application, consideration should be given to providing seating that is resistant to heat absorption from sunlight and impervious to the elements.

Vehicle operators should have clear and unimpaired visibility when running the unit. There should be maximum visibility based upon the functional requirements of the unit. Safety or tempered glass should be used. A roof-mounted stationary window is recommended to add visibility when driving in close proximity to aircraft.

Any vehicle equipped with a windshield should be provided with a powered windshield wiper. On enclosed-cab vehicles, the windshield should be provided with a defogger and/or defroster.

On open-cab vehicles with passenger seats, a hip guard should be provided on the outside edge of the outside seats.

Open-cab vehicles should be provided with slip-resistant materials.

Seats should be provided with a backrest that affords proper lumbar support for the operator.

Door handles, latches and hinges should be rugged, positive, and easily maintainable, and should conform to international safety standards.

Horns should be located so that operators can reach them conveniently.

Rear-view and side-view mirrors should be installed as required.

Hoisting/lifting

The maximum capacity of all hoisting equipment (including chains, cables, etc.) should be displayed in a readily visible location.

Fail-safe devices to prevent any free falling of materials should be provided.

Maintenance safety locks should be provided, and their use required.

A fail-safe means of lowering or disconnecting materials should be provided in case of a malfunction.

Vehicle working surfaces

Work platforms and walkways should have a slip-resistant surface.

Work platforms and walkways should have adequate handrail and/or guardrail protection and toeboards.

Stairs, ramps, and the rungs, cleats, or treads on ladders should be designed to support the minimum working load and be labeled accordingly.

Maneuverability

Wheels on all equipment should be of a type and size that will not cause undue wear to the surfaces over which they will normally operate. For nonmotorized equipment, the wheel type and size should minimize the push/pull efforts of the employees handling the equipment.

All wheels should be equipped with fenders or other suitable devices to protect personnel and materials on the vehicles from spray or splashes.

A power-steering system should be considered for all vehicles.

Exhaust systems

The surfaces and the discharge of exhaust systems should be located so that they will not expose employees to injury or materials to damage in normal operation. Noise levels generated by exhaust systems/engines should not exceed allowable limits.

General

Equipment having an effect on public health (such as food, drinking water and lavatory service equipment) should meet appropriate environmental standards for the country in which the equipment will be operated.

Vehicles carrying materials that could subject the operator and/or occupants to injury in the event of a load shift should be equipped with attach points for load restraint.

All lamps, reflectors and/or reflective tapes should be recessed or otherwise protected against damage by impact. They should meet international motor vehicle standards, as well as local airport requirements. If backup/reversing lights and audible warnings are provided, they should be automatically controlled by the shift selector when it is placed in reverse gear.

All ground support equipment – whether licensed for on-the-road use or not – shall be equipped with clearance lights, reflectors or reflective tape on both sides of the vehicle.

Any towed unit – singular or in train – shall be provided with either clearance lights, reflectors, or reflective tape on both sides and the rear.

The tongues or tow-bars of units should be designed to minimize exposure to pinch points during coupling.

Cushioning devices should be installed on all equipment intended to be used in direct contact with, or in close proximity to, aircraft to minimize aircraft damage. Such cushioning devices should be tubular or D-section bumpers of suitable length, outside diameter, and wall thickness.

Equipment for use in hangars or around aircraft servicing (such as an engine change unit) should comply with NFPA, national and local provisions that deal specifically with aircraft hangars and servicing. (See References.)

GROUND EQUIPMENT MAINTENANCE AND OPERATION

This section highlights some basic procedures to be followed in maintaining and operating ground support equipment.

Maintenance

The required number of operating hours and/or time period for scheduled maintenance and overhaul of all units and their components should be obtained from the vendor or manufacturer.

Equipment components and systems requiring routine and frequent inspections and maintenance should be readily accessible.

Whenever possible, major assemblies and components should be disconnected easily and removed from the equipment. The vendor should specify any assemblies and/or components requiring special procedures and/or tools to perform routine and nonroutine maintenance and overhaul.

All doors and panels should be provided with securing devices to retain them in the open or closed position. These devices should be capable of withstanding jet blast or ambient winds. They should be installed so that the doors, when open, do not create an injury hazard.

All parts subject to high operating temperatures and all sprockets, gears, belts, and other moving parts should be suitably guarded if operating and/or maintenance personnel might accidentally contact them.

All ground support equipment should be inspected at the beginning of the work shift. All items needing correction should be noted, and supervisors or other appropriate authorities should be informed so they can schedule repairs.

Checklists should be made to aid maintenance inspections. For nonmotorized equipment, a checklist should include:

- Ease of operation of the equipment
- Condition of tires and wheels
- Cracks in panels and framework
- Broken welds or missing fasteners
- Condition of hitches
- Condition of hoses, bumpers, fittings and step treads

Checklists for motorized equipment should include

all items for nonmotorized vehicles listed above, as well as the following:

- Condition of grounds and power cables
- Condition of guards and guardrails
- Operation of lights, horns, mirrors and alarm, if so equipped
- Operation of foot and parking brakes
- Pedal travel on brakes and clutch
- Engine, instrument and windshield wiper tests
- Operational damage
- All fluid levels

Equipment that fails any of the above should be scheduled for repair or removed from service (Figure 10-1).

Fluid levels and lubrication schedules on ground equipment should be maintained as specified by the manufacturer. Any other specific checks or service called for by the equipment manufacturer's manual also should be added to the checklist.

Operation

Both motorized and nonmotorized ground support equipment should be parked in designated areas at the terminals between flights. The parking brakes on the equipment should be set, and the gear selector placed in NEUTRAL or PARK position, unless otherwise specified by company policy. Chocking of equipment wheels is highly recommended.

When servicing the aircraft, equipment should be placed in preassigned positions at the aircraft. These positions should be previously specified and illustrated on printed positioning guides (Figure 10-2). A positioning guide should be provided for each type of aircraft the equipment is to service.

When equipment must be either towed or pushed (towing is desirable), it should be performed by qualified personnel only.

Aircraft and pedestrians always should have the right of way. Ground support equipment should never move across the path of boarding or disembarking passengers or taxiing aircraft. Adequate separation

distance should be maintained when crossing behind taxiing aircraft to avoid jet blast (consider a minimum of two aircraft lengths).

Operation of equipment under any part of the aircraft should be clearly defined for type of aircraft and GSE.

No equipment should be permitted to contact an aircraft unless the unit is specifically required to do so to perform its function. If the unit is required to contact the aircraft, auto-leveling aircraft devices should be considered.

As soon as a vehicle is positioned at the aircraft, the gear selector should be placed in the "neutral" or "park" position, the parking brake set, the wheel(s) chocked and the engine turned off (if not required for the function of the unit).

Ground support equipment always should be maintained in good repair. Units in need of repair should be reported immediately by the user to his or her supervisor. The unit must be sent to the appropriate workshop for repair as soon as possible.

An "out of service" tag, to be attached by the operator, is suggested for all units in need of repair. The tags should list the following:

- Type and number of the unit
- Time removed from service
- Reason for removal from service
- Signature of the person removing the unit from service

The "out of service" tag should remain on the vehicle until the vehicle is returned to service by maintenance personnel. The tag should be affixed in a location where an operator will notice it, to prevent inadvertent use of a unit in need of repair.

Riders should not be allowed on any equipment where appropriate seating is not provided.

Vehicles operated away from the airport must have a valid highway license plate(s) and/or special permits.

Monthly Motorized Equipment Safety Checklist

Date _____ Location _____ By _____

CODE: ✓=OK   R=Repair   Y=Yellow Tag   RT=Red Tag

NWA TAG #	EQUIPMENT TYPE	GLASS	NON-SKID		WIND-SHIELD WIPERS	LIGHTS				TIRES				HORN	BRAKES INCLUD-ING EMER-GENCY	STEER-ING & CON-TROLS	GUARD-RAILS, GUIDES, PALLET STOPS	EX-HAUST SYS-TEMS	OTHER REMARKS
			FLOOR	BRAKE	CLUTCH		HEAD	TAIL	TURN	STOP	RF	LF	RR	LR					

DISTRIBUTION: WHITE—Local Plant Maintenance CANARY—Local File

Figure 10-1. The form above is a safety inspection checklist for motorized automotive equipment.

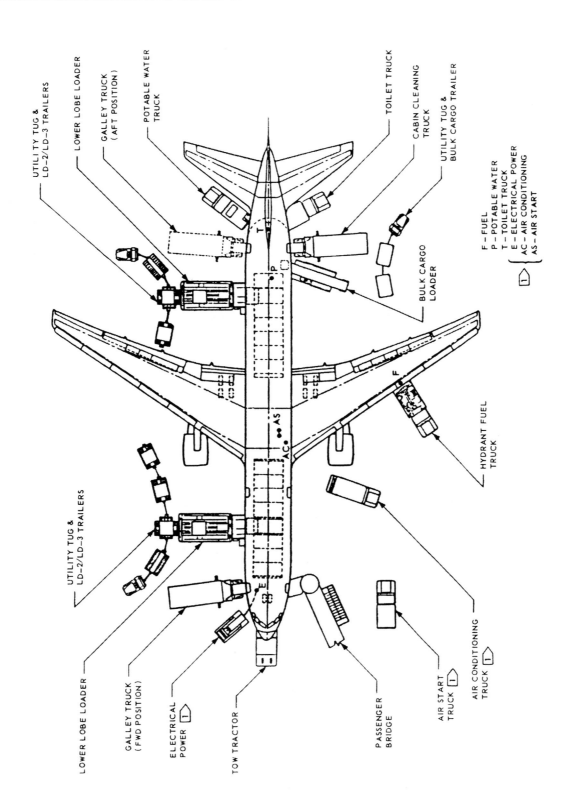

Figure 10-2. For each type of aircraft, a positioning guide specifies where servicing equipment should be located.

PERSONNEL CONSIDERATIONS

This section is about the duties of those who operate ground support equipment. It covers the issues of their qualifications and the training they should receive.

Personnel qualifications

Only qualified personnel, including contractor personnel, should be permitted to operate equipment. They should be trained to operate and handle the equipment for optimum functioning of the unit. Operators should be tested by a qualified instructor and deemed competent. Periodic renewal of a unit operator's competency should be considered. Records should be maintained for all equipment operators.

Strict supervision should be maintained to make sure only qualified operators handle equipment.

Operators should possess a valid driver's license in the jurisdiction or country in which they operate. In addition, special permits may be required by local airport authorities.

Personnel responsibilities

When an operator's vision is restricted in critical areas (such as when positioning certain equipment at an aircraft and removing it from the aircraft), it should be a unit operator's responsibility to use a guide during positioning and/or removal.

No vehicle (motorized or nonmotorized) should carry riders unless such a practice is authorized by company procedures and/or the responsible supervisor. Even then, riders should not be permitted on any unit unless an approved seat and/or approved seating/standing area with handrails or restraints is provided.

Operators should be responsible for the safe operation of the unit they handle. Before using any vehicle, operators should make sure:

• The area is clear of other equipment, personnel and anything else that might interfere with safe operation.

• Cable, hoses and other accessories are properly stowed.
• Doors are closed and secured.
• Horns, lights, windshield wipers and washers operate properly.
• The unit has adequate fuel, oil, water and other required operating fluids. (The above can vary by location based upon company and station policy regarding assignment of responsible groups to maintain and check these items.)

Operators of GSE should observe maximum speed limits. In areas away from the terminal or hangars, the speed limits are usually established by local authorities. When a GSE is approaching or leaving an aircraft or is inside a hangar, terminal or building, the prescribed speed limit usually is a maximum of 8 km/h (5 mph). Speed should be reduced when heavy traffic, bad weather or other adverse conditions exist.

Vehicles approaching an aircraft should be stopped while they are still a safe distance from the aircraft to make sure their brakes are serviceable.

• Vehicles equipped with manual transmissions should approach the aircraft in the lowest gear ratio.
• Vehicles equipped with automatic transmissions should approach the aircraft in the lowest gear range.

Equipment should not approach an aircraft that has its engines running unless it has been specifically cleared into the area by the ground personnel directing the operation.

Training programs

A training program should be established to make sure all operators are taught safe procedures and practices to handle the vehicle(s) they will be required to operate. Where possible, simulations using mock-ups of aircraft doors and loading docks should be provided as part of the training.

It is recommended that the appropriate training manual and procedures for each piece of equipment be used.

During training, special emphasis should be placed on:

- Preuse equipment check
- Defensive driving (driving to save lives, money, and time in spite of conditions around you and the actions of others)
- Working with guide personnel
- Operating in emergencies
- Inclement weather operations

Vehicle operators should be tested for the following skills and behaviors:

- Parking
- Backing
- Turning
- Positioning
- Use of hand and/or turn signals
- Stopping and starting
- Observing clearances

Signaling

For the safe movement of ground support equipment, operators, as well as those who guide them, should be trained in the use of hand signals. It is important that the guide never stand between and directly in front of the approaching vehicle and aircraft. This is to prevent being struck.

The use of approved guide signals is recommended. In the event the operator does not understand the signal, loses sight of the guide or communication is hampered in any way, the operator should immediately stop until the discrepancy is remedied. Conversely, when the guide believes the operator is not following given signals, he or she should signal the operator to stop.

CATERING OPERATIONS

Catering service requires materials handling and the operation of mechanized equipment while the aircraft is on the airport ramp.

Food service handling

Several potential hazards for personnel involved in food service handling should be noted:

- Jamming fingers while opening or closing galley doors
- Tripping or falling over galley units set in narrow galleys or aisles
- Bumping head on open galley compartment doors
- Jamming fingers while inserting or removing ovens, food units and tray carriers
- Burning hands and fingers on ovens
- Getting scratches and cuts from jagged metal edges while handling metal galley units that need repair
- Spraining back due to improper lifting procedures while handling galley units
- Falling out of galley doorways after hi-lift food trucks or other means of access have been removed
- Falling off hi-lift trucks or other means of access
- Lowering bodies of hi-lift trucks without warning (This can injure personnel and damage equipment.)
- Galley units falling or being blown out of aircraft doorways or food trucks
- Falling after attempting to board hi-lift trucks or other such vehicles by unauthorized methods
- Getting scratched as a result of handling broken glass or cutlery in the galley

Operational guidelines

Food service vehicles comprise a considerable part of the ground support equipment that converges on an aircraft during the loading/unloading operation. While food service vehicles usually have special features not found on the typical service vehicle, they should be operated according to the general instructions outlined earlier in this chapter.

Because many of these vehicles are operated by non-airline employees working on a contract basis, it is most important that such operators are required to comply with the same procedures as the airline's own employees. The following safety guidelines outline the procedures needed to safely operate and maintain these vehicles.

Throughout this section, the terms "helper" and "guide" are used interchangeably. For those companies that provide both a driver and a helper on a truck, the helper will be the guide. If only a driver is provided, then when a guide is required to help position a vehicle against an aircraft safely, a member of the ground service crew must serve as the guide.

Qualifications

All drivers and helper/guides should meet the qualifications outlined earlier in this chapter. They should also be trained in the special procedures of each airline for which they provide service.

Supervisors should make periodic observations of a driver's operating habits and performance. Violation of operating procedures should be discussed with the driver on the spot, to make sure he or she is aware of these errors. Violations will be reported to the driver's supervisor at the earliest opportunity.

Drivers are required to check equipment prior to use. Checks should include, but not be limited to, brakes, lights, windshield wipers and horn.

If any of these items are found inoperative, the problem should be reported to the supervisor. The equipment should be recorded for maintenance or tagged "out of service," depending on the seriousness of the problem.

General airport driving rules

The following airport rules should be observed at all times:

- Because of the danger of jet blast, the operator should avoid driving vehicles behind aircraft that have engines running. Look to see if the "anticollision" lights on the aircraft are on. This means that engines are running or are about to be started.
- Aircraft always have the right of way.
- A guide should be used when moving trucks, raising truck bodies or extending platforms when a truck is positioned at the aircraft.
- The guide should stand where he or she can observe the most likely point of contact with aircraft, other GSE, or people (Figure 10-3).
- Drivers always must follow the rule that when they lose sight of the guide, either in the mirror or directly, they must immediately stop.
- The truck must be positioned at an aircraft in such

a manner that no part of the aircraft will contact the truck if the aircraft settles.
- While waiting for aircraft to arrive, trucks should stop only in "authorized" areas and out of the path of passenger loading bridges or other ground support equipment. When a vehicle is parked on the ramp, unattended, the truck body must be fully lowered, parking brake set and engine off.
- Vehicle bodies should not be raised during periods of high wind. Company procedures should be based on the truck manufacturer's specifications.
- Controls for raising bodies must be of the constant-pressure type, and never should be blocked or bypassed.

Loading

The requirements listed in this section, and those that follow, must be met to ensure aircraft are catered safely and properly. The following rules must be observed in loading catering trucks:

- All drivers and helpers should be familiar with galley requirements, to make sure that correct equipment and supplies are put on the trucks.
- Equipment must be handled carefully to avoid personal injury, food spoilage or equipment damage.

Figure 10-3. When the vehicle is about 2.5 m (8 ft) from the aircraft, the guide should stand near the plane to direct the vehicle into its final position.

Photo courtesy of LSG

- All equipment aboard trucks must be stowed securely to prevent shifting in transit.
- Rear doors of trucks must be closed and latched at all times except when personnel are actually entering or leaving the truck body. This is to ensure the safety of personnel, and to prevent anything from blowing out of or falling from trucks.

Positioning trucks at aircraft

Catering vehicles should approach aircraft so operations will be completed properly, minimizing personal injury and potential property damage. Before or upon turning into or toward an aircraft galley door, the catering vehicle should be stopped to test the brakes. (See Airline Catering Safety Guides.)

The operator should then come to a second stop approximately 2.5 m (8 ft) from the aircraft. At this point, the helper will dismount and guide the truck to the final position at the aircraft. Standard hand signals should be used in positioning the truck.

The truck should be positioned so that the bumper will be a short distance from the aircraft skin when the truck is raised to the proper working level. This distance ensures the truck will not come into contact with the aircraft regardless of the movement of the aircraft.

Truck brakes should be set, and the transmission/gear train placed in the correct mode to elevate the truck body. At the same time, the helper should be placing the chock(s) fore and aft of a front wheel of the truck (Figure 10-4).

NOTE: THE USE OF ONE OR TWO CHOCKS IS THE OPTION OF THE COMPANY OPERATING THE TRUCK. WHEN ONLY ONE CHOCK IS PROVIDED, IT WILL BE PLACED IN FRONT OF THE WHEEL.

If at any time during the raising of the truck body, the guide believes the platform is coming too close to the aircraft, he or she should give the "stop" signal. The truck body should be lowered and the truck repositioned to provide sufficient clearance (Figure 10-5).

Figure 10-4. Make sure chocks are in place before the body of the vehicle is raised to service the aircraft.

Photo courtesy of LSG

Raising truck bodies

The helper should assist the driver during the raising operation by closely watching the clearance between the truck and aircraft.

The truck body should be raised only high enough to suit the type of aircraft being catered. When the airline uses a bridge to roll equipment directly into the aircraft, the platform must be at doorsill height.

Truck bodies must not be raised if an outward-opening door of the aircraft is open; a failure in the controls could cause the body to continue to rise too high and damage the door.

Whenever possible, trucks should be positioned so platforms are not under the aircraft door, to preclude any damage if the aircraft settles during loading.

Prior to servicing, the truck platform guardrails (where provided) must be installed to prevent personnel from falling off.

Servicing an aircraft

Before opening the galley door, catering personnel

Figure 10-5. As the body of the service truck is raised, the stabilizers extend automatically.

Photo courtesy of LSG

should knock on it twice to alert anyone inside that it is being opened. Doors should then be opened carefully and moved to the latched position. This will prevent injury or damage from sudden door movement as a result of jet blast or wind.

If, while the door is being opened, it becomes evident that the evacuation slide still is connected, the door should be closed and the airline crew should be instructed to disconnect the slide. An inflating slide can injure personnel on the truck or in the aircraft.

The clearance between aircraft and truck should be checked periodically to make sure no part of the aircraft comes in contact with the truck.

Nothing that can be blown off should be left on the truck platforms. Any equipment that must be staged on the platform should be properly secured.

Aircraft galley equipment will be handled carefully to ensure correct operation in flight. Any unserviceable items should be brought to the attention of the appropriate airline representative.

When catering is completed, catering personnel should check that all galley doors and compartments are properly closed and locked. They should tidy up the galley area and be sure nothing is left behind that could be a problem for the flight attendants.

Leaving the aircraft

Procedures for a truck to leave the aircraft are basically the reverse of positioning the catering truck. The following additional items are required:

- The aircraft door should be closed slowly and carefully; the door should be correctly positioned and latched properly.
- Both the driver and helper should survey the ramp around the truck to make sure the area below is clear of personnel and equipment.
- The driver should lower the body only after he or she has closed the front door, the helper has closed the rear door, and the area is clear. Before leaving the truck body, the driver should make sure the rear door is closed.
- When the driver is seated and has started the engine, the helper will remove the chock from behind the front wheel (where applicable), and will leave the front chock in place.
- The helper or guide will take a position at the rear of the truck-about 3 m (10 ft) behind the truck-where he or she is in full view of the driver in the mirror and where he or she can accurately observe the most critical point of clearance between the truck and the aircraft.
- After the driver has backed the truck approximately 2.5-3 m (8-10 ft), the guide should give the "stop" signal, pick up the chocks, and stow them on the truck. The guide should then return to the rear of the truck and direct the driver away from the aircraft.
- Backing the truck away from the aircraft can be more critical than positioning it. The driver must drive slowly, while observing clearance from the aircraft, and being alert for signals from the guide. If the driver loses sight of the guide, he or she must stop immediately and proceed no farther until visual contact is re-established.
- When the truck is clear of the aircraft and

surrounding equipment, the guide returns to the cab of the truck and the operators proceed to their next assignment.

Special procedures and precautions

Certain aircraft, because of special configurations or means of operation, require additional procedures or precautions in the catering operation. This is particularly true of wide-bodied aircraft because of their height. Such precautions are necessary to protect personnel and to prevent damage to the aircraft.

Procedures must be developed for servicing each type of aircraft, if the aircraft operator does not already have such procedures.

Hazard points that must be addressed and highlighted in the procedures may include:

- Unusually low wing areas of regional jets
- Aircraft cabin doors with power-operated assist should only be operated from the exterior of the aircraft as this will disarm the power assist if inadvertently left in the arm mode.
- In aircraft that are catered by loading complete food units, the catering truck must be positioned to have its platform level with the aircraft door-sill. If a bridge plate is used, it should have a means of being secured so that it does not move when the food containers are transferred from the truck to the aircraft.
- Hi-lift trucks used to service wide-bodied aircraft should have an interlock installed that prevents the body from being raised higher than regular jet aircraft height unless the stabilizers are extended (Figure 10-5).

REFERENCES

Airline Catering Safety Guides

American National Standards Institute:

> *Airline Ground Support Vehicle-Mounted Vertical Lift Devices, Safety Standards for, ANSI A92.7-1998.*
> *Boom-Supported Elevated Work Platforms, ANSI A92.5-1992.*
> *Elevating Work Platforms, Manually Propelled, ANSI A92.3-1990.*
> *Vehicle Mounted Elevating and Rotating Aerial Devices, ANSI A92.2-1990.*
> *Work Platforms, Self-Propelled Elevating, ANSI A92.6-1999.*

Inflight Food Services Association Inc.

International Electrotechnical Commission

International Organization for Standardization

National Safety Council. *Accident Prevention Manual for Industrial Operations, 11th ed.* 1997.

National Fire Protection Association:

> *Aircraft Hangars, NFPA 409, 1997.*
> *Aircraft Maintenance, NFPA 410, 1997.*
> *Powered Industrial Trucks, Including Type Designations, Areas of Use, and Maintenance and Operations, NFPA 505, 1997.*

Society of Automotive Engineers, AGE 2C (Subcommittee)

U.S. Department of Transportation. *Federal Motor Vehicle Safety Standards (FMVSS) in Code of Federal Regulations, Title 49, Part 571.*

INVESTIGATING AIRCRAFT GROUND DAMAGE ACCIDENTS

A systematic approach to investigating aircraft ground damage accidents, which includes the identification of causal factors and implementation of corrective actions, is essential to a good accident prevention program and management system.

DEFINITION OF AIRCRAFT GROUND DAMAGE

Aircraft ground damage accident is defined as an unplanned, undesired event that results in damage to an aircraft while it is on the ground, under control of ground crew members and not during stages of flight under control of the cockpit crew. (e.g., from the time the aircraft leaves an active taxiway under control of cockpit crew until the aircraft has been pushed from the gate and the ground crew has waved off the flight.)

Scope
Aircraft ground damage investigation can potentially involve the expertise of both flight and ground/occu-

pational safety professionals.

Although certain types of ground damage events may fall under the jurisdiction of a variety of governmental agencies, this chapter applies when damage has occurred during certain aircraft movements on the ground. In situations where flight safety professionals become involved, due to specific criteria in the regulations, both flight and ground safety professionals should work cooperatively to investigate the following events:

- Aircraft to or from a gate under the direction of ground crew, whether or not flight crew members and passengers are present
- Aircraft movement from gate to gate
- Aircraft movement from remote location to the gate

The continuum in Figure 11-1 describes the typical involvement of investigative personnel.

Overview
An aggressive accident investigation process includes:

- Identification of the basic causal factors, without placing blame that contributed either directly or indirectly to the accident

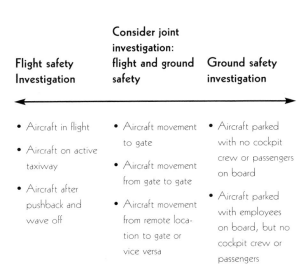

Flight safety Investigation	Consider joint investigation: flight and ground safety	Ground safety investigation
• Aircraft in flight • Aircraft on active taxiway • Aircraft after pushback and wave off	• Aircraft movement to gate • Aircraft movement from gate to gate • Aircraft movement from remote location to gate or vice versa	• Aircraft parked with no cockpit crew or passengers on board • Aircraft parked with employees on board, but no cockpit crew or passengers

Figure 11-1. Typical investigative involvement.

- Recommendations for corrective actions
- Recommendations for improvements of the management system
- Defining an organization's written plan for aircraft ground damage events
- Define coordination with corporate legal departments

When an accident occurs, it usually means something went wrong with the management system: An oversight, an omission or lack of control over the work process allowed the accident to occur. The investigation process must determine not only the direct causes of the event, but also deficiencies in the management system that contributed to the causes.

A good damage accident investigation will uncover a number of contributing factors and subsequent corrective actions. The corrective actions should include both specific measures needed to eliminate or reduce the probability of recurrence of a similar type of accident and the measures necessary to improve the management system.

As the management system improves, overall aircraft ground damage accident rates should also improve. Management must support the investigation process and act on the recommendations that result from the process. Additionally, improved productivity, morale, quality and cost reduction will be achieved.

Effective ground damage accident investigations require strong management commitment and involvement. Investigators must be capable of and knowledgeable about the principles of investigation and interview techniques. They also must have adequate resources available to conduct the investigation. If management does not show strong support, employees may conclude safety and damage accidents are not taken seriously. When a vigorous, objective investigation is conducted and the ensuing recommendations are followed, management commitment is clear.

Role of the safety department
The safety department may play a number of different roles, based on the organization and the severity of damage. Typically, safety department members oversee or monitor the investigation process. Line departments typically may be responsible for collecting information, securing the evidence and supporting the interview processes. Government agencies also may be involved.

The safety department should evaluate the quality of all damage investigations. The evaluation should include an assessment of:

- Accuracy, timeliness and completeness of the report
- Clear description of events leading to the damage event
- Identification and description of all potential causal factors:
  - Failure to follow procedures
  - Human error
  - Mechanical problem
  - Corrective actions that were taken immediately after the event
- Appropriate review and signatures

An accident investigation, if conducted properly, will identify what happened, how it happened and why it happened (or why it could have happened, in the event of a "near miss.") It also will indicate what should be done to prevent recurrences. Safety should play a leading role in establishing the investigation process for the organization and training involved employees in the process for optimum results.

Additionally, the safety department is responsible for collection of information necessary to calculate damage rates, trends, identify problem areas and provide departmental comparisons.

Investigation criteria
All aircraft ground damage accidents must be investigated, regardless of the extent of the actual or potential damage. Assessment of the damage or potential for damage may dictate the extent of the investigation required (See Chapter 2, "Risk Management").

Objectivity/no blame

Investigations should identify causal factors and a plan for preventing repeated accidents. Investigations should not be used for disciplinary or punitive actions. Blaming individuals for damage jeopardizes the investigation's credibility and effectiveness, and usually will reduce the quality and accuracy of the information received. In summary, blaming can result in a failure to identify proper corrective actions. Additionally, a corporate culture that focuses on blame and punishment can result in failure to report future aircraft damage events.

Relevant acts of oversight or acts of omission or commission on the part of employees, supervisors or other management personnel should not be ignored, however. It is critical to determine why someone did or did not follow an established procedure that resulted in aircraft ground damage. Guidelines for managing employee behaviors are identified in Figure 11-2.

Supervisors should have the interest and objectivity needed to carry out good investigations in their own departments. If the safety department encounters an individual who can't or won't be objective, an independent investigation should be completed by the safety department.

Managing behaviors

The makeup and size of the investigative team should be dictated by the seriousness or complexity of the ground damage event. The supervisor, with the help of the involved employees usually investigates minor ground damages. Following the initial assessment and based on the degree of severity, investigation teams can include:

- Employees directly involved with the work process where the damage occurred
- Supervisors
- Technical specialists (airport planners, ergonomists, engineers, consultants, etc.)
- Trainers and others who have familiarity with the operation
- Employee representatives

The team can add more members as the investigation unfolds. The investigative team members should have investigation training and the following characteristics:

- Subject matter expertise
- Regulatory requirements
- Technical knowledge
- Objectivity
- Tactful communication skills
- Good analytical skills
- Curiosity
- Detail oriented
- Rigorous

An aircraft ground damage investigation plan should provide for designation of a team leader and team members. The investigative team leader should be

Errors resulting from behaviors related to system issues	Errors resulting from unintentional* at-risk behaviors	Errors resulting from intentional risk-taking behaviors
Manage through changes in: • Processes • Procedures • Training • Aircraft design • Environment • Ground equipment design • Facility design	*Unintentional: The person did not know their actions would cause a bad outcome. **Manage through:** • Defining at-risk behaviors • Removing incentives for at-risk behaviors • Creating incentives for appropriate behaviors • Increasing situational awareness	Purpose and Knowledge: The person knew their actions would cause a bad outcome. Recklessness: The person knew there was a high probability their actions would cause a bad outcome. **Manage through:** • Disciplinary action

Figure 11-2. Managing behaviors.

designated at the onset of the investigation. The leader should have management status and the authority to accomplish the investigation. Anyone designated to investigate aircraft ground damage must be competently trained and prepared to act promptly.

The leader's duties should include:

- Calling participants together to discuss the event
- Coordinating team activities, including onsite observations
- Assigning tasks and schedules
- Ensuring no potentially useful data source is overlooked
- Keeping interested parties apprised of the progress of the investigation
- Overseeing the preparation and distribution of the final report
- Arranging liaison with employee representatives
- Coordinating with involved department regarding recommended follow-up action items

Components of a written investigation plan

The management system should include a written plan for team investigations. The plan should address the following:

- Identification of the person in charge, who can assign responsibility
- Authority to conduct investigations
- Skills training for investigators
- Prompt notification of team members indicating where and when they should report
- Instructions regarding the personal protective equipment to be worn at the scene
- Provisions for a work area and administrative support
- Transportation and communications needed
- Securing the scene for the duration of the investigation after rescue and damage control are finished
- Procedures/equipment to ensure the observation and recording of fragile, perishable or transient evidence such as instrument readings, control panel settings, weather/environmental conditions, chemical spills, stains, skid marks, work schedules, etc.

- Development and distribution of a comprehensive report
- Tracking method to ensure follow up and compliance

PROCESS FOR AIRCRAFT GROUND DAMAGE ACCIDENT INVESTIGATIONS

Notification

Involved parties need to be made aware of the ground damage accident. A formal notification process must be adopted to ensure all appropriate persons receive timely notification of the accident. Typically for commercial aviation, the supervisor of the involved process notifies the company's operational control center or equivalent, where a paging or "call tree" process has been established. The severity of the accident or damage will dictate the depth of the notification process. Notification call lists must be reviewed periodically or subsequent to major organizational changes to keep the notification lists accurate.

Local airport authorities may need to be contacted, based on the location or magnitude of the accident. It may be necessary to notify the local airport authority, law enforcement, fire/rescue teams or company security. There may be legal requirements to notify government regulators and investigation authorities. Written company policies and procedures should address when this becomes necessary.

Immediate response

The immediate safety and health of employees, passengers and the public must be the main concern following a serious ground damage accident. Activities related to the investigation are secondary. The first response should be to:

- Provide emergency response and first aid
- Secure and isolate the scene
- Identify names of witnesses and other involved parties for interviewing
- Collect transient or perishable evidence
- Determine the extent of damage to equipment

and/or aircraft
- Restore the operation as much as possible

See Appendix A for sample reporting forms.

Collecting facts

The severity of real or potential damage will determine the overall effort involved in the investigations. The investigator(s) should perform tasks relevant to the identification of causal factors. For major ground damage accidents, the investigation team should:

- Visit the scene of the accident immediately, before any evidence is disturbed, whenever possible. When the scene is secured, the investigation is enhanced.
- Take photographs or videos of the scene from a distance, as well as detailed photos. A digital camera is preferred. Since no one can predict in advance which data will be useful, numerous photos should be taken from many different angles and distances. Reference indicators, like rulers, can be very helpful to indicate the basic dimensions of scale. Secure any available videotapes of the scene from airport authorities, as applicable.
- Make accurate and complete sketches or diagrams. These should be made before the scene is disturbed or restored. Showing the path of travel of a vehicle up to the point of impact with an aircraft, for example, can be very helpful.
- Preserve and secure physical evidence from the event. When the investigation reveals an equipment failure or damage, arrangements should be made to either preserve the item as it was found at the scene, or to document any subsequent repairs or modifications.

Types of evidence

It is important to separate facts from opinions, direct evidence from circumstantial evidence and eyewitness accounts from hearsay. Data and information collected during the investigation can be listed in the following categories:

- Hard evidence: Factual information, which cannot

be disputed, such as time and place, written reports and possession of physical evidence.
- Witness statements: Statements gathered from persons who saw the accident occur and from those who came upon the scene immediately afterward. Witness statements can be written or captured through an interview process. (See Appendix B for sample witness report format.)
- Circumstantial evidence: The logical interpretation of facts that leads to a single, but unproven conclusion.

Caution should be used when handling physical evidence such as equipment and tools. It may be necessary to send this evidence to a laboratory for further analysis of failure or fractures if visual assessments are inconclusive. Ensure the item is identified, tagged and secured. It may be necessary to establish a chain of custody to prevent evidence tampering.

Attempts should not be made to reassemble failed or damaged equipment. Such attempts may jeopardize an accurate laboratory analysis. Likewise, disassembly of damaged equipment is not recommended unless the correct reassembly procedure is known. Investigators should determine the necessary analysis and refer the actual analytical work to the experts. (See Appendix C for more information on investigation kits.)

- Interview witnesses and other involved parties as soon as possible. The quality of their statements is highest immediately after the accident. Immediate interviews minimize the possibility that witnesses will subconsciously adjust their stories to fit the interviewers concept of what happened or to protect someone involved. Witnesses should be interviewed individually and in private so their comments do not influence each other.
- Conduct interviews in a manner that prevents apprehension. Skilled, tactful investigators usually get cooperation from employees by eliminating any apprehension they may have about incriminating themselves or others. Witnesses must be convinced

that the investigators are searching to find the cause of the accident and not someone to blame. (See Appendix D for additional information on investigation techniques.)

- Review all sources of potentially useful information. This will avoid unwarranted impression that information actually obtained from witnesses is based on the investigator's own observations or analysis. Documentation of these sources can prove valuable if the investigation is expanded or reopened at a later date. Note any contradictory statements or evidence and attempt to resolve the discrepancies. Examples of information sources are as follows:
- Equipment design specifications
- Drawings
- Operating logs
- Purchasing records
- Written and verbal procedures or instructions
- Risk assessments and previous investigation reports
- Equipment operating manuals
- Maintenance inspection and test records
- Modification or design change records
- Job safety analyses
- Training records
- Job performance records
- Lab tests
- Cockpit voice recordings
- Video surveillance tapes
- Airport authorities
- Police reports

Accident investigation report
The report should include the following:

1. General description of the accident, including the date, time, aircraft number, involved ground equipment, location and name of investigator.
2. Detailed report of the chronological events that occurred.
3. Contributing factors and analysis. Describe what led/contributed to the failures.

   a. Equipment/materials
   b. Work procedures
      - Written
      - Verbal

   c. Environmental

      - Weather
      - Time of day
      - Facility lighting
      - Noise

   d. People

      - Training
      - Supervision
      - Motivation

   e. Human factors:

      - Fatigue
      - Distractions

   f. Organizational influences

      - Culture on the ramp

- Practices or standard operating procedures
- Time pressures

g. Drug or alcohol testing completed, as permitted.

h. Cause and effect diagram of preceding causes, causes and outcomes

i. Recommendations

j. Reference attachments, like written statements, photographs, maintenance records, training records, references to company policies and procedures, etc.

k. Distribution of the report should be noted.

NOTE: IF A FATALITY IS CONNECTED TO AN AIRCRAFT GROUND DAMAGE ACCIDENT, ENSURE THAT THE PROPER AUTHORITIES HAVE BEEN NOTIFIED. INVESTIGATORS SHOULD NOT ATTEMPT TO DETERMINE THE CAUSE OF DEATH. (SEE APPENDIX D FOR EXAMPLES OF AIRCRAFT GROUND DAMAGE INVESTIGATION SUMMARY REPORTS.)

REFERENCES

Air Force Safety Center, Public and Media Affairs Division, Kirtland AFB
National Safety Council

WORKING GROUP MEMBERS

Co-Chairmen:

Terri Weiland – TR Weiland Company, LLC

William R. Jaggi – American Airlines

Members:

Bill Carlyon – Boeing Commercial Aviation

Fred Rose – Air Wisconsin

Jim Stephan – Delta Airlines

Larry Hennemann – Northwest Airlines

David Marquette – Marsh Risk Consulting

Tom Pazell – United States Air Force

Holly Zimmerman – Alaska Airlines

Roland (Jim) Page – Page Safety Solutions

Contributors:

David Dufour – Northwest Airlines

Gerry Pipe – Air Canada

Cathy Hollister – Air Canada

Tracy Sora – Air Canada

Mohn Murphy – Northwest Airlines

Ricardo Moreno – Allied Caterers Ltd./BWIA

Margi Evenson – America West Airlines

APPENDIX A: SAMPLE EVENT REPORTING FORM; SAMPLE SUMMARY REPORT FORMAT

Report #_____

INSTRUCTIONS: This form is to be filled out in the event of aircraft, equipment, or facility damage. All Accidents should be investigated to determine cause and means to prevent recurrence. **PLEASE PRINT.** Attach additional sheets if necessary. Do not leave any portion blank. Indicate "N/A" where necessary.

General Information

Station/Department	Date and Time of Accident/ Discovery	Investigator's Name/Title (PRINT)
	/ / MO / DA / YR Military Time	

Damage To (check all that apply)

☐ Aircraft   ☐ Facility
☐ Equipment   ☐ Public Property
☐ Other

Other Party Damage (if applicable)

☐ Damage to other party's aircraft/equipment/facility
☐ Damage to HP aircraft/equipment/facility by other party

Other Party

Name

Address

Telephone

Conditions at Scene of Mishap

If conditions at scene are unknown, check here ☐

Visibility (check one)

☐ Good   ☐ Marginal ☐ Poor

Obstructions to Visibility:

Lighting (check one)

☐ Day   ☐ Night   ☐ Dusk

Artificial Light (check one)

☐ Good   ☐ Marginal ☐ Poor   ☐ None

Surface Conditions (check all that apply)

☐ Dry   ☐ Snow   ☐ Ice   ☐ Wet   ☐ Other

Weather (check all that apply)

☐ Clear   ☐ Snow   ☐ Hail   ☐ Fog   ☐ Sleet
☐ Windy   ☐ Rain

Individuals Involved (Identify individuals involved, including witnesses, and have them fill out a Witness Report.)

Name and Position Last, First	Position	Company Name	Badge #	Phone #	Date of Hire	Hours on Duty	Consecutive Days Worked

Type of Release	Type of Substance		Type of Container		Released Quantity:
☐ Spill ☐ Leak	☐ Fuel ☐ Oil ☐ Hydraulic Fluid	☐ Lav Waste ☐ De-icing Fluid ☐ Other (describe)	☐ A/C Fuel Tank ☐ Storage unit ☐ Mobile unit; ie. Truck etc.	☐ Drum ☐ Other	_____ gal Size: X

*Did spill reach storm drain? (attach documentation)

Describe clean up: equipment and personnel used, *outcome/completion? (attach documentation)

A/C - Flight Information

A/C No

A/C Type

Flight No
   o Non-Flight

Was Flight?

☐ Canceled

☐ Delayed

Length of Delay

Scheduled Ground Time

Actual Ground Time

Bag/Cargo Load

No. Passengers Rebooked

☐ On AWA

☐ On OAL

Damaged Areas (check all that apply)

☐ Fuselage   ☐ Cowling
   ☐ Forward   ☐ L
   ☐ Aft   ☐ R
☐ Tail   ☐ Radome
☐ Entry Door   ☐ Pitot Tube
☐ Galley Door   ☐ AOA Vane
   ☐ Forward
   ☐ Aft
☐ Rear Left
☐ Rear Right
☐ Cargo Door   ☐ Antenna
   ☐ Forward
   ☐ Rear
☐ Wing   ☐ Interior
   ☐ L ☐ R
☐ Wing Tip ☐ GPU Panel
   ☐ L ☐ R
☐ Horizontal Stabilizer   ☐ Lav Panel
☐ Fuel Panel/Cap
☐ Spoiler
☐ Flap
☐ Fairing
☐ MainGear
☐ Nose Gear

☐ Engine
   1 ☐ 2 ☐ 3 ☐ 4 ☐

Airworthiness

Was repair required to maintain

Airworthiness?   Y / N

Phase of Operation

☐ In Flight   ☐ Ground Servicing
☐ Taxiing   (lavs, etc.)
☐ RON   ☐ Unknown
☐ Towing   ☐ Other
☐ Hangar
☐ Pushback
☐ Fueling
☐ Deicing
☐ Catering
☐ Loading/Unloading

Contributing Factors

Equipment (check all that apply)

☐ Malfunction (verified)
☐ Design Problem
☐ Not Guarded
☐ Not Maintained
☐ Inappropriate for Task
☐ Inaccessible
☐ Using Broken Equipment
"Inop" or ""Caution" Tagged?
   ☐ Yes ☐ No
☐ Other
☐ N/A
☐ Unknown

Environment (check all that apply)

☐ Congestion
☐ Lighting
☐ Weather
☐ Noise
☐ Cleanliness of Work Area
☐ Other
☐ N/A
☐ Unknown

Human Performance (check all that apply)

☐ Distraction
☐ Peer Pressure
☐ Boredom/Complacency
☐ Unfamiliar with Task
☐ Fatigue
☐ No Procedure in Place
☐ Procedure Not Enforced
☐ Procedure Not Used
☐ Procedure Used Not Effective
☐ Time Constraints
☐ Lack of/poor Communication
☐ Physical Limitations of
☐ Employee
☐ Horseplay
☐ Improper Delegation of Task
☐ Other
☐ N/A
☐ Unknown

Description of Event

Describe the sequence of events as you determine them. Please provide details to describe categories checked under "Contributing Factors."

-Did substance reach storm drain?

-How and by whom was the spill cleaned up?

-What was the disposition of the clean up materials?

NOTE: Attach witness Reports, BOI documentation, aircraft diagrams and photographs.

Persons Injured?   Yes   No   If yes, include Safety Injury Report:

Describe Damage: (dimensions; severity)

Suggested Actions to Prevent Recurrence (if damage occurred at your location)

Distribution of Form

1. FAX TO CORPORATE SAFETY AS SOON AS COMPLETED

2. COMAIL ORIGINAL TO CORPORATE SAFETY WITHIN 24 HOURS

3. KEEP COPY ON FILE AT YOUR STATION

AIRCRAFT MISHAP SITE DIAGRAM	
STATION:	FLIGHT #:
TIME OF REPORT:	AIRCRAFT #:
PREPARER'S NAME:	

DIRECTIONS: Draw a diagram of the aircraft, which notes the location and extent of damage, location of equipment or personnel involved and any other obstacles or factors.

NOTE: PLEASE USE THIS FORM FOR ALL AIRCRAFT TYPES

APPENDIX B: WITNESS STATEMENT REPORT – SAMPLE FORM

WITNESS REPORT

Date of Report _____   Flight # _____

Reporting Station _____   Aircraft # _____

(Please mark the appropriate box)

☐ Aircraft Damage   ☐ On The Job Injury   ☐ Equipment or Facility Damage

When did this happen? _____
_____
_____

Where did this happen? _____
_____
_____

Who was involved? _____
_____
_____

What happened? _____
_____
_____
_____
_____

Why did this happen? _____
_____
_____

Report Submitted By:_____

Employee #: _____

Manager / Supervisor:

APPENDIX C: ACCIDENT INVESTIGATION KIT

Investigation Kits typically contain the following items to facilitate the investigation process.

- ☐ Camera (film and digital)
- ☐ Caution tape
- ☐ Investigation forms
- ☐ Tape measure
- ☐ Compass
- ☐ Tape recorder
- ☐ Paper and pens, including graph paper
- ☐ PPE:
  - ☐ Safety glasses
  - ☐ Hearing protective devices
  - ☐ Work gloves
  - ☐ Disposable vinyl gloves
- ☐ Flashlight
- ☐ Magnifying glass
- ☐ Investigation procedure and the following additional documents:
  - ☐ Emergency contact list
  - ☐ Interviewing techniques
  - ☐ Flow charts
  - ☐ Investigation tips
  - ☐ Checklists

APPENDIX D: SAMPLE INVESTIGATION TECHNIQUE TIPS

1. Gain trust and let witnesses know they are providing a service to the company. They should understand that they are being helpful and you are appreciative of their cooperation.

2. Don't press too hard for answers; be pleasant and positive so they help you. If they feel threatened, they may purposely mislead you or omit information.

3. Interview in a non-threatening, neutral location free of interruptions and distractions.

4. Try to remain on the same level as the witness. Have chairs at an equal height and situate yourself close to the witness as you would if you were having a casual conversation. Don't separate yourself with a large imposing desk.

5. Reassure the witness that you are not looking to place blame or point the finger at any employee - emphasize learning for improvement and prevention.

6. Let the witness recount the events before, during and immediately after the incident without interruption.

7. When the witness is finished, ask questions to clarify what you heard and restate their account in your own words to confirm the story.

8. Ask open questions that will not be leading and will not put words into their mouths. For example, ask "Did you notice anything unusual upon arriving at the aircraft?" rather than "Were the chocks placed incorrectly?"

9. Ask how they would prevent a repeat accident. The employee may be an expert on operating the equipment or completing the task and may have valuable suggestions for improvement.

10. Ask Who, What, When, Why, How, and Where.

WHO...
- saw the damage event occur?
- was working with him/her?
- instructed, trained, assigned?
- else was involved?
- can help prevent recurrence?

HOW...
- did the damage occur?
- could the damage have been avoided?
- could the supervisor have helped?
- did others contribute to the damage?

WHAT...
- was the type of damage?
- task was being performed?
- were employees told to do?
- equipment or tools were being used?
- precautions were necessary?

- problem or question was encountered?
- did witnesses see?
- procedures were followed/not followed?
- procedures are needed?
- what was done just before, just after the incident?

WHY...
- can be asked up to five times following any of the other questions to help get to the root cause.

WHEN...
- did the damage occur?
- was the task begun?
- was the task assigned?
- were the problems realized?
- was the supervisor notified?

WHERE...
- did the damage occur?
- was the witness at the time of the incident?
- was the supervisor at the time?
- were other people involved at the time?
- else might this occur?
- was the equipment positioned?

APPENDIX E: SUMMARY REPORT

Sample Investigation Summary Report

SERIOUS/MAJOR OCCURRENCE

        Occurrence Date:

        Occurrence Time:

        Aircraft No.:

        Ground Equipment No.:

        Location:

        Report Date:

        Estimated Costs: Material:

                               Labor:

                               Total: (Direct Costs)

        Submitted by:

GENERAL DESCRIPTION OF OCCURRENCE:

DETAILED REPORT:

*

*

CONTRIBUTING FACTORS AND ANALYSIS: What led to the failures? (Include Cause and Effect Diagram)

Preceding Cause(s)	Cause:	Outcome
	• Failure to follow procedures	
	• Human error	
	• Mechanical failure	

Describe the type of aircraft damage or potential damage.

Describe what contributed to the cause of the outcome.	Describe the cause or causes. Add boxes as needed.

Describe other real or potential outcomes of the event.

Work Procedures: (Written, Verbal)

Environment: (Weather, Time of Day, Facility lighting or noise)

People: (Training, Supervision, Motivation)

Equipment/Material:

Human Factors: (Fatigue, Distractions)

Organizational Influences:

Drug Testing completed? Yes _____   No _____

DOCUMENTS AVAILABLE IN OH&S INCLUDE:

DISTRIBUTION:

Original:

Copy to:

PREVENTING INADVERTENT
SLIDE DEPLOYMENTS

CHAPTER 12

The inadvertent deployment of aircraft emergency evacuation slides is an aviation industry problem. Hundreds of slides are deployed each year. Each deployment is costly due to the expense of repacking the slides, flight delays and cancellations. Each deployment also has the potential to cause serious injury.

DEFINITIONS

Inadvertent slide deployment

For the purpose of this chapter an inadvertent slide deployment is the non-intentional deployment of an aircraft emergency evacuation slide. This includes passenger, main and (B747) upper deck doors, off-the-wing slides and tail slides as well as tail cones.

Slides deployed during an emergency evacuation, functional tests or when there is an equipment malfunction will not be considered an inadvertent slide deployment.

Deployed. A slide will be considered deployed if the door was in an armed condition when opened and:

- The slide was pulled from its container
- The door assist bottle was discharged (on aircraft so equipped)

The slide does not have to inflate to be considered deployed by this definition.

Malfunction. If a slide is deployed while opening a door due to a malfunction of the door or slide mechanism (broken linkage, seized parts, etc.), it will not be considered a slide deployment.

Slide deployments during maintenance. If a slide is deployed while maintenance or an inspection is being performed on a door or slide, it will be charged as a slide deployment if proper disarming procedures were not followed. If the slide deployed due to a malfunction, it will not be charged as a slide deployment.

Emergency power assist doors. Wide body aircraft and Boeing 757 aircraft have power assist passenger doors. These doors have power assist bottles or spring systems that, when activated for emergency egress, forcibly open the door and the slide. These doors are armed (put in the emergency mode) by switches or levers, and when operating properly, will not deploy slides when the door is opened from outside the aircraft.

Mechanical doors. Narrow body aircraft have slides that must be physically attached to the aircraft floor to arm them for an emergency. The slide is activated when the door is opened and the slide is pulled from its storage pack because it is attached to the aircraft floor. Slides on these doors will deploy if the slide is attached to the aircraft floor (armed) and someone opens the door from the inside or the outside.

Girt bar. The girt bar is the device connected to aircraft emergency slides that attaches the slide to brackets at the door threshold when emergency egress is needed. With the girt bar attached to the floor, the slide is pulled from its container when the door is opened. If the slide is then able to fall a number of inches out the door it will inflate. The girt bar then holds the slide in place at the door so that passengers and crew may use it for egress.

Discussion

Inadvertent slide deployments occur when aircraft doors are opened while the doors are in the emergency or "armed" position in non-emergency situations.

Doors may be in the armed position for several reasons:

- The door was not disarmed after the aircraft arrived at a gate. It is normal procedure to disarm the door upon arrival.
- The door was disarmed by the flight crew on arrival but was armed in error at a later time by someone else.
- The person opening the door inadvertantly armed the door as they opened the door.
- The door may have been armed in preparation for departure.

Impact of slide deployments

Several costs are associated with slide deployment. Some of the major ones are:

- Cost of injuries that may occur due to personnel or passengers being struck by a deploying slide or an automatic door
- Cost of inspection and re-packing the slide
- Cost of the flight delays or cancellations
- Lost revenue from flying with blocked seats if another slide is not available locally
- Maintenance costs for damages that may occur

An industry average for inspection and re-packing of a blown slide (defined as one that is not damaged) is $6,000. All other costs vary greatly depending on the circumstances of the event. Slide deployments identified as aircraft ground damage are currently the number one cause of ground damage to aircraft.

Identifying causes of inadvertent slide deployments

- Distractions that cause a person to miss disarming the door
- Distractions while opening an armed door
- Untrained personnel opening the door
- Maintenance personnel not following proper disarming or safety procedures prior to performing maintenance on doors.
- Confusing signage
- Door design

Identifying the aircraft with the most slide deployments

Surveys done in 2002 indicate that Airbus and Boeing 757 aircraft are most the most susceptible to inadvertent slide deployments. The Lockheed L1011 was least susceptible. The reason appears to be due to the different types of levers, switches, and indicators that arm, disarm, or open doors.

Who is blowing the slides?

- Cabin cleaners
- Catering
- Flight attendants and flight crews
- Passenger service agents

- Maintenance
- Passengers

Industry practices to prevent inadvertent slide deployments:

- Procedures
- Awareness programs
- Engineering changes
- Incident investigations
- Training

PROCEDURES

General procedures

1. On-through flights require that doors on narrow-bodied aircraft be opened (cracked) from the inside by in-flight service prior to anyone attempting to open the doors from outside.
2. Implement very specific procedures for each step of the operation:

   - Who opens doors
   - When they are opened and closed
   - Who disarms doors and when

3. Require doors on all automatic door aircraft to be opened from the outside only, whenever possible.
4. Restrict door operation authority to only those trained.
5. Restrict contract service door opening authority to contract service supervision.
6. Report near-misses; when a door is found armed, report it so the cause can be tracked.

In-flight service (check, check, and cross check) procedures

1. Have flight attendants "own" a door, that is, one flight attendant works the same door from the start of the flight to the end. This will help ensure there is never any doubt about who should be arming, disarming, or "cracking" or opening a door.
2. Perform cross check whenever arming or disarming doors, have flight attendants arm or disarm their door then "cross check" (look across) to ensure the other flight attendant performed the same function. Then attendants should report to an on-board leader that they have accomplished these tasks (the arming or disarming and the cross checking).
3. Have the rear-most flight attendants re-check all doors as they leave the aircraft.

Maintenance procedures

1. Standardize procedures for repairs or service on all aircraft doors, and ensure concise verbiage is used with appropriate cautions at critical steps. For example: "Verify slide disarming procedures have been accomplished," or "Warning: Failure to disarm escape system prior to removing escape hatches will cause accidental deployment of slide and possible injury to personnel or equipment."
2. Create a "Door Operation Orientation and Slide Deployment Prevention" training package for each fleet type—there are lots of aircraft and lots of differences. Require all mechanics to review the procedure, and make it part of recurrent and new hire training.
3. If a fleet type or a particular door is causing problems, require additional training for employees working on those fleets or doors.

Contract services procedures

1. Make it policy that only trained personnel operate doors. It may be helpful to require that only supervision be allowed to operate doors.
2. Monitor and enforce procedures, stop the shortcuts that seem to always be involved with deployments. One example is cabin service opening up doors to throw trash bags to the ramp. Loading trash onto high-lift trucks already positioned by the access door, or taking it out through the loading bridge or down aft stairs if those were the means of access, will eliminate someone opening an additional door and potentially deploying a slide.
3. Require door operation familiarization training for all that come on aircraft even if their job doesn't require operating the doors. Many slides have been blown because some one wanted to help out. When possible teach it in a language they are most comfortable with.
4. Review procedures to try to find ways to minimize

the number of people that open doors and the number of times doors are opened.

5. Have supervision check that all doors are disarmed before work crews board aircraft. This is especially important with large cleaning crews. Ensure the supervisors know not to try and disarm doors if they find any armed. They need to leave it alone and report it. Make it a policy that only flight attendants and mechanics may disarm doors.

Awareness

1. Conduct incident conference calls as soon as possible after an incident. If you can get the local supervision and upper management together, it can be an effective way to get to root cause and to increase awareness. (See section on incident investigation for more details on conference calls)

2. Use newsletters and daily shift briefing to pass along details of all slide deployments as soon as they happen. The briefings should include identification of the root cause and emphasize corrective action.

3. Post photos of your deployments – they can sometimes be real eye-catchers (See Figures 12-1 through 12-3).

Engineering

1. If trends are noted, such as slides being deployed on only one aircraft type, look for possible system design issues.

2. Examine the type of warning labels, arming levers and pins used on all doors. Consult with engineering for possible changes on doors that are causing problems.

3. Talk with other airlines with similar equipment – they may have found a solution already.

4. Work with aircraft manufactures – they may be able to show you different designs that may help prevent deployments.

5. Work on improving signage – in some cases this may be a simple but effective solution.

Some examples of engineering changes:

- Use a strap across door window to show door is still armed.
- Add pins to lock some arming levers in safe.

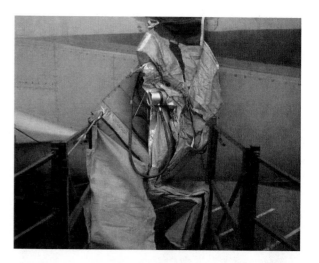

Figure 12-1 through 12- 3. Posting photos of deployments can increase awareness and serve as eye-catchers.

- Add flaps that hang down over the arming lever saying the door is armed when it is.
- Modify the girt bar brackets on the door to hold the girt bar more securely to the door when the girt bar is stowed. This is to help prevent the girt bar from falling off door and causing a slide deployment.

Incident investigation

1. Treat all inadvertent slide deployments as aircraft ground damage and fully investigate to find the root cause.
2. Hold conference calls with select personnel involved in the incident as soon as possible after the incident. This can be an effective tool to start gathering the facts about an incident. A call soon after an event also keeps the incident investigation from being put as a low-priority item, and tends to get everyone's attention as to its importance. The format for conference calls should be part of your incident investigation procedures. Here are some guidelines:

   - Create a list of the personnel who need to participate (high-level personnel participating in conference call – especially one that fall on weekends or after hours – tends to set a tone that the company takes the incident seriously). Some believe not including the actually person(s) involved and just including supervision helps keep the call from becoming an inquisition.
   - Require the call to be held within 24 hours of the incident. Some companies require it within three hours – even it falls at 2 a.m. They believe this puts the importance of the call at a high level.
   - Require the call to be no longer than 30 minutes.
   - Develop a set of basic fact-finding questions to use for all calls. This way the call leader is steered toward fact-finding and kept away from blame-placing.

3. Look for trends and develop corrective actions based on your investigation. Do not let the investigation be just a piece of paper. Perform follow-up to ensure corrective actions have been implemented

Training

Training is one of the key elements in any slide-deployment program. Who conducts the training depends on your organization. Formal classroom training, OJT, home study, and videos all have a place. A combination of all the above seems to be the most effective. Some organizations use door mock-ups to train employees; other employees must be trained on actual doors. Aircraft manufacturers sometimes have training material, but it may not be specific to the type of arming and disarming methods your aircraft doors have.

1. All personnel authorized to operate doors must be trained on the methods to identify an armed door, and trained in the proper way to open and close the doors they will operate.

   - The training should be step-by-step. All models and modifications of doors that will be encountered should be included.
   - The training should include cautions about not attempting to operate doors that the employee is not familiar with or has not been trained on.

2. All personnel who work near or around aircraft doors should be trained on door operation – whether or not their job requires them to open or close doors. If they are not authorized to operate aircraft doors they should be trained to not touch a door or open or close one.

MAINTENANCE OPERATIONS, AIRCRAFT DOOR OPERATION, ORIENTATION, AND EMERGENCY SLIDE DEPLOYMENT PREVENTION

Objectives

- Establish a "Don't touch if you don't know" culture surrounding the operation of aircraft doors and emergency hatches in the passenger cabin of aircraft during maintenance.
- Communicate the dangers and costs associated with improperly operating these systems.
- Show the basic operating differences between fleet types.
- Identify the doors that remain armed in the maintenance environment

Responsibilities

It is the responsibility of maintenance supervision to ensure all personnel operating aircraft doors know and adhere to procedures and have had all required training.

NOTE: PERSONNEL CAUSING DAMAGE OR INJURIES DUE TO IMPROPER OPERATION OF AIRCRAFT DOORS OR HATCHES WILL BE HELD ACCOUNTABLE.

Training

All maintenance personnel are require to have the following training and attend the safety briefings outlined.

• All mechanics shall receive "Door Operation Orientation and Emergency Slide Deployment Prevention" training prior to working on aircraft.
• All mechanics working on the B767 fleet are also required to have "Off-Wing Slide Arm/Disarm" training (the off-wing slide system on this aircraft has been a particular problem).
• The safety business plan should require that sections of the "Door Operation Orientation and Emergency Egress Deployment Orientation" program be covered throughout the year in safety briefings to maintain awareness of the hazards of aircraft doors.

General issues

Accidental deployment of emergency slides is a major safety concern. Accidental or inadvertent slide deployments can result in:

• Injuries
• Aircraft or equipment damage
• Flight delays
• Disruption of service
• Flight cancellations
• Passenger inconvenience and dissatisfaction

All of these are very costly. All slide deployments cause maintenance expense just for inspection and re-packing even if they do not cause any of the above. This maintenance cost is around $6,000 and that is if the slide is not damaged. Repairs if slides are damaged can greatly increase this amount. Add the maintenance costs to the flight-disruption costs (lost revenue, hotels and meals for passengers, etc.) and the average cost associated with a slide deployment can be put at approximately $40,000.

Mid-cabin egress hatches and doors

Emergency egress from the middle of the passenger cabin is accomplished by a variety of means.

B767-400 series aircraft mid-cabin egress is accomplished through an emergency exit door located on each side of the fuselage, just aft the wing (Figure 12-4). The slide cannot be disarmed without removing the slide bustle and the slide pack from the door. Operating the emergency handle to open this door will cause the following:

• The emergency door is hinged at the bottom of the door and will rise as the handle is moved. The door then will fall outboard and down against the side of the fuselage.
• As the door opens the girt of the slide is pulled, and the slide will deploy.

This system remains armed during all base maintenance and line maintenance operations unless the maintenance manual specifically calls to disarm the system. Anyone working in the area of the deployment could be seriously injured or killed.

Figure 12-4.

Photo courtesy of Delta Airlines

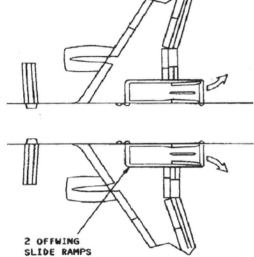

2 OFFWING SLIDE RAMPS

Figure 12-5.

Photos courtesy of Delta Airlines

B757 aircraft mid-cabin egress is accomplished through two emergency exit hatches located on each side of the fuselage, directly over the wing. Removal of either of these hatches without disarming the system properly will result in the following:

- The slide compartment will blow open in the wing fillet area, just aft of the hatch locations.
- The slide will deploy outboard and off the wing trailing edge.

The off-wing slide system is deactivated during all heavy service visits. For all other scheduled and unscheduled maintenance performed by line maintenance and base maintenance, the doors are armed unless disarming is called out by the maintenance manual. Anyone working in the area of the deployment could be seriously injured or killed.

B767 200 & 300 series aircraft mid-cabin egress is accomplished through single or dual emergency exit hatches located on each side of the fuselage, directly over the wing (Figure 12-6). Removal of either of these hatches without disarming the system properly will result in the following:

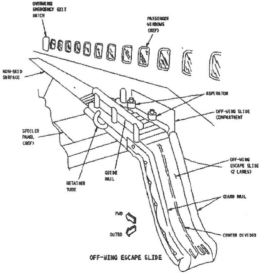

Figure 12-6.

Photo courtesy of Delta Airlines

- The slide compartment will blow open in the wing fillet area, just aft of the hatch locations.
- The inboard spoilers will blow down to allow the slide to deploy without obstructions.
- The slide will deploy outboard and off the wing trailing edge.

The off-wing slide system is deactivated during all heavy service visits. For all other scheduled and unscheduled maintenance performed by line maintenance and base maintenance, the doors are armed unless the maintenance manual specifically requires the system to be disarmed. Anyone working in the area of the deployment could be seriously injured or killed.

MD88, MD90, B727 and B737 mid-cabin egress is accomplished through two exit hatches located on each side of the fuselage, directly over the wing.

These aircraft wings are close enough to the ground so that off-wing slide assemblies are not required to provide egress from the wing.

The trailing edge flaps are lowered, and straps are put in place to guide the passengers aft where they slide down the lowered flap assemblies – off the trailing edge of the wing.

MD11, B777 and L1011 mid-cabin egress is accomplished through additional passenger and service doors located in the area just forward and aft of the wing, or directly over the wing.

These doors are equipped with slide systems and should be treated the same as the doors covered in the following section on passenger and service doors.

Passenger and service doors

B757 passenger/service door configuration is shown. Door configuration for each fleet type can be found in the aircraft's maintenance manual.

Emergency egress from the forward and aft ends of the passenger cabin is accomplished through slide systems incorporated into the operation of the pas-

senger and service doors located in these areas.

Because these doors serve multiple purposes, loading and unloading of passengers and equipment as well as emergency egress. The egress systems are designed to allow them to be disarmed quickly and easily when procedures are followed correctly.

Failure to properly disarm the emergency egress system at any passenger or service door could result in the deployment of the slide pack for that door.

Door operation and the emergency systems the airline's fleet varies greatly. Personnel should not operate any passenger or service door with which they are not familiar.

Personnel must always follow appropriate maintenance procedures. Heed all warnings and follow all procedures for disarming and rearming the system when required for the maintenance being performed.

Basic differences in passenger/service door operation for aircraft types.

B727, B737, MD88 and MD90 passenger/ service doors. The emergency egress system must be disarmed manually from inside the aircraft. The normal procedure is for the aircraft flight attendants to disarm (manually remove girt bar from floor brackets and stow) the door upon aircraft arrival at a gate (See Figure 12-7).

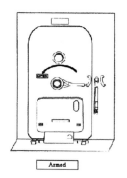
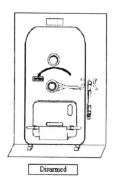

Armed   Disarmed

Figure 12-7.

Caution: This is not always done. Always assume the door is armed. Check prior to operating or working on door. If you have not been trained or are not familiar with door do not touch get help.

Opening doors from outside: Doors on these aircraft cannot be operated from the exterior handle (outside) safely if the door was not first disarmed (girt bar removed from floor and properly stowed) from inside the aircraft.

Opening doors from inside: Ensure the girt bar is removed from floor brackets and properly stowed before operating the door from inside the aircraft.

MD88 and MD90 aft entry door (ventral stairway). Because the MD88 and MD90 have only one aft entry/service door located in the L2 position, these aircraft are required to have a slide system incorporated along with the aft vertical stairway.

The normal ground operation of this door is to open and close it using the handle located just above the door-mounted flight attendant seat. The center section of the pad should be raised into the "up" position, covering the emergency handle. When flight attendants prepare the cabin for departure, they lower the flap to cover the open/close handle on the door, exposing the emergency handle. If the door is opened using the emergency handle the following will occur:

- The tail cone will be jettisoned.
- The vertical egress slide will deploy.

B757, B767, B777, MD11 and L1011 passenger/service doors. The emergency egress system disarming procedure is to disarm the doors on arrival of aircraft at gates. This is done by flight attendants.

Caution: This is not always done. Always assume door is armed, check prior to operating or working on doors. If you have not been trained on or are not familiar with door do not touch door get help.

Opening doors from outside: When the door is opened properly with the exterior handle (from out-

Figure 12-8. B757/B767-200 Series Aircraft A typical B757 online today.

side), the emergency egress system is disarmed even if it had been left armed, and the door can be rotated to the open position safely. Caution: If the door had been left armed and it is opened from the outside the door will disarm, but when closed from the outside it will rearm itself again if it has not been disarmed from inside aircraft.

Opening doors from inside: When preparing to open the door from inside the aircraft, you must first ensure the arm/disarm handle is in the "disarm" position prior to opening the door.

Caution: Passenger/service doors on these fleet types are equipped with a mechanical system to assist in the opening of the door when the door is armed. These systems are in place to ensure that the door opens fully in the event of an emergency, and there is sufficient clearance for the egress slide to deploy. The assist system will open door with sufficient force so that anyone or anything in the path of the door when the assist function is activated could be seriously injured or damaged.

DETERRENT DEVICES FOR MID-CABIN OFF-WING EXIT HATCHES

Because the mid-cabin hatches on aircraft with off-wing slides remained armed while aircraft is at gates or undergoing maintenance at gates or in maintenance facilities (except while in heavy service), they are particularly hazardous. The following is a description of devices that may be used during maintenance to remind personnel these hatches are armed.

For view of the B767 ETOPS emergency hatches and overhead bins see Figure 12-9.

EXAMPLE OF A CATERING PROGRAM

Overview
The inadvertent deployment of aircraft emergency slides is a dangerous, costly problem. Slides are unintentionally deployed for several reasons: Lack of knowledge of system, in a hurry, not thinking, not following procedures are some key reasons.

Objectives
- Establish a "Don't touch if you don't know" culture surrounding the operation of aircraft doors.
- Communicate the dangers and costs associated with improperly operating aircraft doors.
- Outline some simple steps that will help prevent inadvertent slide deployments.
- Identify some hazards that are particular to catering.

Responsibilities
- It is the responsibility of catering supervision to ensure all personnel operating aircraft doors know

Figure 12-9. Here we have a view of a domestic B767-300.

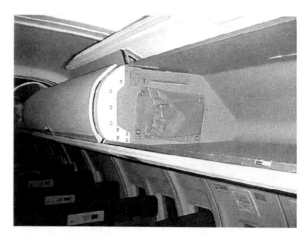

Figure 12-10. Pouch in position in a B757 above the over wing emergency escape hatch.

and adhere to procedures, and have had all required training.
- It is the responsibility of catering supervision to ensure all personnel working around aircraft doors (even if they are not authorized to operate doors) know the hazards associated with them.
- It is the responsibility of catering supervision to ensure personnel are following all the proper procedures.

NOTE: PERSONNEL NOT FOLLOWING PROCEDURES WILL BE HELD ACCOUNTABLE.

Personnel causing damage or injuries due to improper operation of aircraft doors will be held accountable.

Procedures
NOTE: ENSURE ONLY DOOR TRAINED PERSONNEL WORK AIRCRAFT

Ensure that only personnel who are trained in door opening and closing open or close doors.

CAUTION: TREAT ALL AIRCRAFT DOORS AS IF THEY ARE ARMED UNTIL YOU HAVE VERIFIED THAT THEY ARE DISARMED.

1. If you are unfamiliar with or have not been trained on an aircraft door, do not attempt to operate it.
2. Open aircraft doors from the outside only.

CAUTION: ON LIVE FLIGHTS IF LOADING BRIDGE IS NOT TO AIRCRAFT (EITHER IT HAS BEEN REMOVED FOR DEPARTURE OR

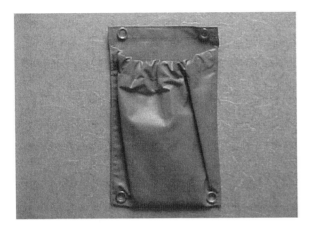

Figure 12-11. This life vest pouch found on B767-200 aircraft is our preference for the use to hold the Emergency Exit Handle Cover Plates and should hold costs down.

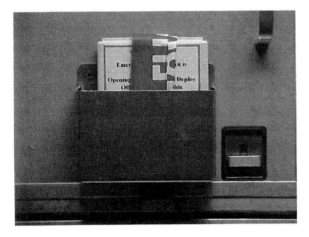

Figure 12-12. Napkin holders found on various aircraft can also be used to protect the proposed streamer handle cover plates from possible baggage damage.

HAS NOT BEEN PLACED TO AIRCRAFT YET ON ARRIVAL) BE AWARE THAT AIRCRAFT DOORS ARE LIKELY TO BE ARMED.

3. Before opening the door verify that the door area is clear of obstructions, the door will clear the trucks rails, the truck platform is at the correct height, and the area by the door inside the aircraft is clear.
4. If it is a live flight ensure in-flight service is ready by knocking on the door and watching for a "thumbs-up" from a flight attendant.
5. Open the door slowly. Stop if you feel an obstruction; if the door is not operating normally, stop

and get help from a flight attendant.
6. If you find a door armed, do not attempt to disarm – guard and get help.
7. Do not open doors for ventilation.
8. Do not open a door unless a stand, loading bridge, or vehicle is at the door.
9. Avoid opening doors unnecessarily.

Training

All catering personnel that work on aircraft are required to have the following training and to attend the safety briefings outlined.

- All caterers shall receive "Door Operation Orientation and Emergency Slide Deployment Prevention" training prior to working on aircraft.
- All caterers shall receive additional on-the-job training during their first three weeks working on an aircraft. This training shall include having a mentor for the three-week period. This mentor will assist the newly assigned person with any questions and will be responsible for observing that person while performing duties at least once a week during the OJT period.
- All caterers that work aircraft are required to attend "Door Operation Orientation and Emergency Slide Deployment Prevention" recurrent training annually.
- Slide deployment prevention shall be a topic at least quarterly at monthly safety meetings.
- The "Door Operation Orientation and Emergency Slide Deployment Prevention" program should be covered throughout the year in safety briefings to maintain awareness of the hazards of aircraft doors.

Incident investigation

When a slide deployment occurs that involves catering personnel, an investigation should be started immediately. Supervision should obtain written statements from all involved, and the circumstances of the event should be outlined in a report. The objective of the investigation shall be to find the root cause – not place blame.

The following are the steps to follow:

1. Report the time and place of the incident to headquarters.
2. Obtain written statements from all involved.

3. Take pictures of the scene, if possible.

4. Hold a conference call within 12 hours of the incident with personnel involved in the incident, their supervision, and the safety department. Remember to keep the call to 30 minutes and keep it to fact-finding and away from placing blame.

5. Look for trends, and develop corrective actions based on the facts.

6. Produce a report within five business days of the incident.

7. Ensure employees at all locations are made aware of the details of the incident and the finding of the investigation.

8. Follow up to ensure all action items of the investigation are implemented.

EXAMPLE OF A CABIN SERVICE PROGRAM

Overview

The inadvertent deployment of aircraft emergency slides is a dangerous, costly problem. Slides are unintentionally deployed for several reasons: Lack of knowledge of system, in a hurry, not thinking, and not following procedures are some key reasons.

Objectives

- Establish a "Don't touch if you don't know" culture surrounding the operation of aircraft doors.
- Communicate the dangers and costs associated with improperly operating aircraft doors.
- Outline some simple steps that will help prevent inadvertent slide deployments.
- Identify some hazards that are particular to cabin service.

Responsibilities

- It is the responsibility of cabin service supervision to ensure all personnel operating aircraft doors know and adhere to procedures and have had all required training.
- It is the responsibility of cabin service supervision to ensure all personnel working around aircraft doors (even if they are not authorized to operate doors) know the hazards associated with them.
- It is the responsibility of cabin service supervision to ensure personnel are following all the proper procedures.

NOTE: PERSONNEL NOT FOLLOWING PROCEDURES WILL BE HELD ACCOUNTABLE.

Personnel causing damage or injuries due to improper operation of aircraft doors will be held accountable.

Procedures

NOTE: ENSURE THAT ONLY PERSONNEL WHO ARE TRAINED IN DOOR OPENING AND CLOSING OPEN OR CLOSE DOORS. ALSO ENSURE THAT ALL PERSONNEL WHO WORK ON AIRCRAFT HAVE RECEIVED DOOR FAMILIARIZATION TRAINING EVEN IF THEY ARE NOT TO OPEN DOORS.

CAUTION: TREAT ALL AIRCRAFT DOORS AS IF THEY ARE ARMED UNTIL YOU HAVE VERIFIED THAT THEY ARE DISARMED.

1. If you are unfamiliar with or have not been trained on an aircraft door, do not attempt to operate it.

2. Open aircraft doors from the outside whenever possible – avoid opening doors from the inside.

3. Before opening the door verify that door area is clear of obstructions, the door will clear the trucks rails, the truck platform is at the correct height, and the area by the door inside the aircraft is clear.

4. If it is a live flight ensure in-flight service is ready by knocking on the door and watching for a "thumbs-up" from a flight attendant.

5. Open the door slowly. Stop if you feel an obstruction; if the door is not operating normally, stop and get help from a flight attendant.

6. If you find a door armed, do not attempt to disarm – guard and get help.

7. Do not open doors for ventilation.

8. Do not open a door unless a stand, loading bridge, or vehicle is at the door.

9. Avoid opening doors unnecessarily.

Training

All cabin service personnel who work on aircraft are required to have the following training and to attend the safety briefings outlined.

- All agents shall receive "Door Operation Orientation and Emergency Slide Deployment Prevention" training prior to working on aircraft.
- All agents shall receive additional on-the-job train-

ing during their first three weeks working on an aircraft. This training shall include having a mentor for the three-week period. This mentor will assist the newly assigned person with any questions and will be responsible for observing that person while performing duties at least once a week during the OJT period.

- All agents who work aircraft are required to attend "Door Operation Orientation and Emergency Slide Deployment Prevention" recurrent training annually.
- Slide deployment prevention shall be a topic at least quarterly at monthly safety meetings.
- The "Door Operation Orientation and Emergency Egress Deployment Orientation" program should be covered throughout the year in safety briefing to maintain awareness of the hazards of aircraft doors.

Incident investigation

When a slide deployment occurs that involves cabin service personnel, an investigation should be started immediately. Supervision should obtain written statements from all involved, and the circumstances of the event should be outlined in a report. The objective of the investigation shall be to find the root cause – not place blame.

The following are the steps to follow:

1. Report the time and place of the incident to headquarters.
2. Obtain written statements from all involved.
3. Take pictures of the scene, if possible.
4. Hold a conference call within 12 hours of the incident with personnel involved in the incident, their supervision, and the safety department. Remember to keep the call to 30 minutes and to keep it to fact-finding and away from placing blame.
5. Look for trends, and develop corrective actions based on the facts.
6. Produce a report within five business days of the incident.
7. Ensure employees at all locations are made aware of the details of the incident and the findings of the investigation.
8. Follow up to ensure all action items of the investigation are implemented.

Example of an In-Flight Service Program

Overview

The inadvertent deployment of aircraft emergency slides is a dangerous, costly problem. Slides are unintentionally deployed for several reasons: Lack of knowledge of system, in a hurry, not thinking, distractions (passenger needing help etc.), not following procedures are all key.

Objectives

- Reinforce a procedural based culture surrounding the operation of aircraft doors.
- Communicate the dangers and costs associated with improperly operating aircraft doors.
- Outline some simple steps that will help prevent inadvertent slide deployments.
- Identify some hazards that are particular to in-flight Service.

Responsibilities

Each flight attendant is responsible for knowing the procedures for arming, disarming, opening, and closing aircraft doors.

Procedures

CAUTION: TREAT ALL AIRCRAFT DOORS AS IF THEY ARE ARMED UNTIL YOU HAVE VERIFIED THAT THEY ARE DISARMED.

NOTE: FLIGHT ATTENDANT PROCEDURES FOR OPERATING DOORS OF EACH AIRCRAFT TYPE ARE FOUND IN THE IN-FLIGHT SERVICE ON-BOARD MANUAL.

Below is a listing of some general items for informational purposes only.

On arrival

1. When a loading bridge or mobile stair approach aircraft, disarm the door you are assigned to and check (cross check) that the person disarming door across from you has disarmed theirs.
2. Report the status of your door and the fact that you cross-checked to the onboard leader.
3. If you are the door safety flight attendant, (flight attendant stationed in the cabin's rear most position) check that all doors on the aircraft have been

disarmed as you leave the aircraft. Report the completion of your check to onboard leader.
4. The onboard leader should confirm that all door checks have been completed.

NOTE: ON NARROW BODY AIRCRAFT THE FLIGHT ATTENDANT DISARMS ALL DOORS AND CRACKS THEM OPEN. THIS IS SO THAT THE PEOPLE SERVICING THE AIRCRAFT CAN TELL THAT THE DOOR HAS BEEN DISARMED AND THEY CAN OPEN DOOR.

On all wide body aircraft the flight should not be opening the door from the inside, though they may assist passenger service on the forward passenger door once passenger service has started to move mechanical doors.

Training
• Slide deployment awareness should be a routine item at all safety meetings.

Incident investigation
When a slide deployment occurs that involves in-flight service personnel an investigation should be started immediately. Supervision should obtain written statements from all involved and the circumstances of the event should be outlined in a report. The objective of the investigation should be to find the root cause – not place blame.

The following are the steps to follow:

1. Report the time and place of the incident to headquarters.
2. Obtain written statements from all involved.
3. Look for trends and develop corrective actions based on the facts.
4. Produce a report within five business days of the event.
5. Ensure employees at all locations are made aware of the details of the event and the findings of the investigation.
6. Follow up to ensure all action items of the investigation are implemented.

WORKING GROUP MEMBERS

Chairman:

Jim Stephan – Delta Airlines

Members:

Tom Dyce – Air Canada

Everson I. Pereira – Singapore Airlines

Willy Pfister – SP Techinics, TEB

PERSONAL PROTECTIVE
EQUIPMENT

CHAPTER 13

Personal protective equipment
Eye and face protection
Hearing loss prevention
Head protection
Respiratory protection
Skin protection
Special work clothing
References

Employees need, and are required to use, protective equipment when hazards or environmental conditions call for it. Chemical hazards, radiological hazards and mechanical irritants are capable of causing injury or impairments to bodily function. They can affect the eyes, ears, respiratory system, head, face, hands, trunk or feet through absorption, inhalation or physical contact.

Directors, managers and supervisors should evaluate their operations to determine what substances or operations are potentially hazardous. Correcting potential hazards may involve one or more of the following (the order of this list is an U.S. Occupational Safety and Health Administration recommendation and may not be accepted worldwide):

1. Elimination
2. Substitution
3. Engineering controls
4. Administrative controls
5. Personal protective equipment.

These control measures were discussed in Chapter 3, "Hazardous Materials and Harmful Physical Agents."

PERSONAL PROTECTIVE EQUIPMENT

Certain general principles should govern the selection of personal protective equipment. The equipment should be appropriate for the environment in which it is being used. For example, it would not be appropriate to provide acid handlers with the same kind of apron worn by a welder. Nor would one furnish a welder with a rubber and plastic outfit suitable for acid handlers.

Personal protective equipment and clothing should fit the user, do its job well and cause minimum discomfort. All three factors should be considered if the equipment is to provide maximum effectiveness.

It is the employer's responsibility to assess the workplace to determine if hazards are present, or are likely to be present, that necessitate the use of PPE. This workplace hazard assessment will be verified in written certification that identifies the workplace evaluated. The employer also is responsible for selecting the type of PPE to protect the employee, and must provide training on the use and limitations of PPE.

Eye and face protection

All work areas and specific jobs that create flying particles, harmful liquid splashes, or harmful rays should be identified. All persons exposed to these hazards should be required to wear proper eye or face protection (Figure 13-1).

In some cases, an entire work area (such as one in which a number of machine tools or chemical tanks are located) should be classified as a mandatory eye protection area. In other cases, eye protection may be required only for the employee performing a particular operation (such as refilling a battery or checking hydraulic fluid in a landing gear).

Losing an eye may be costly under workers' compensation laws and losing a skilled operator's services can result in additional cost, but neither can compare to the cost suffered by the injured employee.

Operators who need corrective lenses may use prescription safety glasses or goggles.

Employees working with chemicals should avoid wearing contact lenses. Standard safety goggles that cover ordinary glasses may also be worn, but they require cups that are deep and wide enough to accommodate the entire eyeglass area. Face shields normally are designed to provide protection for the face (that is, the front part of the head including forehead, cheeks, nose, mouth, chin and neck) but not

Figure 13-1. Employee protective eyewear should be suitable to the task being performed.

Photo courtesy of Northwest Airlines

for the eyes. Therefore, such devices should be worn over suitable, basic eye protection.

Applicable standards are in the ANSI (American National Standards Institute) Occupational and Educational Eye and Face Protection. (See References.) Note that occasionally inappropriate or deficient personal protective devices are sold as standard protective eyewear. To avoid mistakes in purchasing, be sure that the current ANSI standard is the one with which the manufacturer or supplier complies.

Hearing loss prevention (OSHA's Noise Standard 29 CFR 1910.95)

Every year, according to the National Institute for Occupational Safety and Health (NIOSH), approximately 30 million workers in the United States are exposed to hazardous levels of noise in the work environment. This makes noise one of the most common health problems in American workplaces. Exposure to high levels of noise may cause hear-

ing loss, cause stress, reduce productivity and contribute to accidents. However, noise-induced hearing loss can be reduced or eliminated by implementing engineering controls and hearing conservation programs. The OSHA noise standard 29 CFR 1910.95 was promulgated to reduce and/or eliminate noise hazards in general industry (manufacturing, utilities and service sectors), and does not cover the construction or the oil and gas well drilling and servicing industries.

The standard has undergone several changes and amendments since it was first introduced, and was most recently amended to reflect new recordkeeping requirements that took effect Jan. 1, 2003.

OSHA's hearing conservation standard mandates that all feasible administrative and/or engineering controls must first be considered to reduce noise levels below 90 dB. Only when controls fail to reduce noise below 90 dB must exposed operators wear hearing protectors sufficient to reduce noise exposure below 90 dB 8-hour time weighted average (TWA). OSHA 29 CFR 1910.95 stipulates that operators be enrolled in a hearing conservation program when occupational noise exposure equals or exceeds an 8-hour TWA of 85 dB measured on the A scale (slow response) or a noise dose of 50 percent. For purposes of the hearing conservation program, employee noise exposure shall be established in accordance with appendix A and Table G-16a of the standard, and without regard to any personal protective equipment attenuation.

The following is an overview of primary compliance issues related to OSHA 29 CFR 1910.95. While this overview describes most compliance directives, it suggests you refer to the OSHA 29 CFR document printed in the *Federal Register:*

• Adequate and effective hearing protection must be made available to employees exposed to 85 dB 8-hour TWA.
• Annual hearing testing should be made available to employees exposed to 85 dB 8-hour TWA. Hearing testing and protection must be made available free

of charge to the employees. Employees should be given the opportunity to select their hearing protectors from a variety of suitable hearing protectors.

- Annual protective training must be conducted for employees exposed to 85 dB 8-hour TWA. Training content must include proper use, insertion, care of hearing protectors, and an explanation of hearing procedures and results.

- Baseline audiograms for new hires should be obtained within the period of six months following hire date (12 months allowed when a mobile hearing test service is utilized). If baseline audiograms are established more than six months after the employee's first exposure at or above the action level, employees should wear hearing protectors for any period exceeding six months after first exposure until the baseline audiogram is obtained. After the initial baseline audiogram, the employer must obtain a new audiogram annually for each employee exposed at or above an 8-hour time-weighted average of 85 dB.

- Operators exhibiting a standard threshold shift must be notified of that shift in writing within 21 days of determination; retests may be conducted within 30 days of determination. A standard threshold shift (STS) is defined as "a change in hearing threshold of 10 dB or more averaged at 2,000, 3,000 and 4,000 Hz in either ear as compared to the reference or baseline audiogram." When establishing a standard threshold shift, a correction in the annual audiogram may be made for the contribution of aging (presbycusis) to the change in the hearing level in accordance with the procedure described in Appendix F of the standard. All threshold shifts of 10 dB from an employee's initial hearing test (baseline audiogram) that have resulted in a total 25 dB level of hearing above audiometric zero must be recorded on the OSHA 300 Log within 6 days of the shift notification. The old criteria required recording 25 dB shifts. If the shift is later shown by retest to be repealed, the employer may "line-off" the affected operator's name from the log. An operator may also be lined-off the log if a physician determines that the shift is medically related (non-occupational).

If it is determined that a standard threshold shift is work-related, employees not using hearing protectors shall be fitted with, trained in their use and care, and required to wear hearing protectors that attenuate noise exposure below a dose of 85 dB 8-hour TWA. Employees already using hearing protectors should be refitted and retrained in the use of hearing protectors. It may also be necessary to provide hearing protection that offers greater noise attenuation.

- Operators with "problem" audiograms must be referred to a physician or audiologist for professional review.
- When a physician determines a worker's ear-related medical condition is occupational, the employer shall cover the cost of the referral visit.
- Noise surveys, exposure measurements, measurement equipment calibrations and hearing test records must be maintained and made available to employees, former employees, representatives designated by the individual employee, and the Assistant Secretary (OSHA).
- Noise exposure measurement records should be kept on file for two years.
- Audiometric test records should be retained for the duration of the affected employee's employment.
- If the employer ceases to do business, the employer must transfer to the successor employer all records required to be maintained by the noise standard.
- The noise standard must be posted in a conspicuous place (like a notice board in the employee break room) for employee viewing.

Noise monitoring

A sampling strategy shall be designed to identify employees for inclusion in the Hearing Conservation Program and to enable selection of appropriate hearing protection. Where circumstances such as high operator mobility, significant variations in sound level, or a significant component of impulse noise make area monitoring generally inappropriate, the employer shall use representative personal sampling unless the company can show that area sampling produces equivalent results.

Continuous, intermittent and impulsive sound levels from 80 dBA to 130 dBA shall be integrated into the noise measurements. Instruments used to measure employee noise exposure shall be calibrated to ensure measurement accuracy.

Monitoring shall be repeated whenever a change in production, process, equipment or controls increase noise exposures to the extent that:

• Additional operators may be exposed at or above the action level.

  or

• The attenuation provided by hearing protectors being used by operators may be rendered inadequate to properly protect the employee.

The company shall notify each employee exposed at or above an 8-hour TWA of 85 dBA. Additionally, the company shall provide all affected employees or their representatives with an opportunity to observe any noise measurements conducted.

Hearing protection
It is estimated by the U.S. Department of Labor that 35 million to 40 million operators are exposed to potentially hazardous noise in the workplace.

Noise-induced hearing loss is mostly preventable. It is preventable through the proper use of hearing protectors, strict control of exposure time, and operator education and training.

The first line of defense against the detrimental effects of occupational noise is administrative and/or engineering controls. The second line of defense is through operator training, education, and motivation. The last and perhaps most vital line of defense is adequate and correct usage of hearing protection.

A number of professional organizations are available as a resource to in the specialty of occupational hearing conservation:

• National Hearing Conservation Association (NHCA)
• Council for the Accreditation of Occupational Hearing Conservation (CAOHC)
• American Speech-Language Hearing Association (ASHA)
• National Institute for Occupational Safety and Health (NIOSH)
• U.S. Occupational Safety and Health Administration (OSHA)

Head protection
Employees who work in areas where there is a potential for injury to the head should be required to wear head protection that complies with ANSI Z89.1. (See References.) Protective helmets designed to reduce electrical shock hazard should be worn by each affected employee when near exposed electrical conductors which could contact the head (OSHA 1910.135). Some operations permit the use of protective headgear known as a bump cap, a thin-shelled, lightweight plastic cap. Originally conceived for aircraft operators performing operations within the confines of a fuselage, it is not standard protective headgear. There are no specifications covering bump caps. Authorities warn that, although bump caps are used for some applications, they are not a substitute for hard hats.

Respiratory protection
Engineering measures, such as ventilation, should be used to control occupational diseases caused by breathing air contaminants. Air may be fouled by harmful dusts, fogs, fumes, gases, smoke, sprays or vapors. When engineering controls are not feasible, appropriate respirators should be used. Respiratory protective devices fall into three classes: air-purifying, air-supplied and self-contained breathing apparatus. Only personal protective equipment that conforms to published standards should be used. Its use should be under control of someone knowledgeable on the subject. (See ANSI Z88.2 and the NIOSH standards in References.)

Air-purifying devices. Air-purifying devices remove contaminants from the atmosphere. They can be used only in atmospheres containing sufficient oxygen to sustain life (at least 16 percent of volume at sea level) and within the specified concentration limitations of the specific device. Air-purifying devices operate in one or both of these ways: Various chemicals remove specific gases and vapors, and/or mechanical filters remove particulate matter. The useful life of an air-purifying device is dependent upon contaminant concentration, wearer-breathing volume, and air-purifying-medium capacity.

Mechanical filter respirators provide respiratory protection against particulate matter such as nonvolatile dusts, mists, or metal fumes. Selecting the appropriate respirator is based on type, toxicity and size of particulate matter.

Chemical cartridge respirators provide respiratory protection against certain gases and vapors in concentrations not in excess of 0.1 percent.

Combination chemical cartridge and mechanical filter respirators provide respiratory protection where exposure is both gaseous and particulate.

It should be recognized that chemical cartridges have a limited storage time. They should not be used after the expiration date. Storage of unsealed cartridges is not permitted, as it may be possible that they may be contaminated.

Gas masks provide respiratory protection against certain specific gases and vapors in concentrations up to 2 percent or as specified on the canister label, as well as against particulate matter.

Air-supplied devices. Air-supplied devices deliver breathing air through a supply hose connected to the wearer's facepiece. Air delivered must be free of contaminants (including oil from the compressor). The air source must be located in clean air and automatically monitored. With the exception of hose masks with blowers, these devices should be used only in atmospheres not immediately dangerous to life. Basic types of supplied-air devices are air-line respirators with constant flow or demand flow, and hose masks with or without blower.

Self-contained breathing apparatus. Self-contained breathing apparatus provide complete breathing protection for various periods of time based on the amount of breathing air or oxygen supplied and the breathing demand of the wearer. Basic self-contained breathing apparatus include the oxygen cylinder rebreathing type and the chemical oxygen rebreathing type (self-generating or oxygen demand). Although proper selection depends upon the application and service time required, complete respiratory protection is provided and the problem of proper protection for a specified hazard is eliminated.

How and when to use each type. Selecting proper respiratory equipment should be based on the following:

- Identify substance or substances that mandate protection.
- Know the hazards and significant properties of each substance, as related on the Material Safety Data Sheet (MSDS).
- Determine exposure conditions and air contaminant levels, and test for oxygen levels.
- Ascertain whether any human capabilities and limitations are essential to safe use.
- Fit the respirator to each operator carefully, and instruct operators in its use and upkeep.
- In some countries medical surveillance is mandatory for operators wearing respiratory equipment.

One limitation applies to all respiratory protective equipment. Certain gases can harm the body by means other than through the respiratory tract. For example, ammonia, in concentrations of approximately 3 percent or higher, can cause skin burns. To avoid that possibility, impervious clothing may be required in addition to respiratory protection.

Similarly, protective clothing should be worn when appreciable amounts of gases such as hydrocyanic acid are present. Hydrocyanic acid – a vapor at slightly above room temperature – is capable of penetrating skin and causing systemic poisoning. To do so, however, concentrations must be considerably higher than those required for poisoning through the respiratory tract.

A continuous, documented program of respirator inspection, cleaning and disinfecting, repair, and storage – as well as training for all employees who use respiratory protection – should be in place.

Skin protection

Employees exposed to skin irritants need to keep high standards of personal cleanliness. Good hygiene helps reduce dermatitis and should be monitored. Employees should be instructed how and when to wash. The type of soap they use is important, because even a generally mild soap may cause skin irritation. The choice of soap is best left to the medical department or others who are qualified.

Skin creams and lotions offer some protection against various chemicals, solvents, and fuels used in the aviation industry. However, their value is limited. If other protective devices can be used, creams and lotions only should be used to back up basic protection. Water and soluble products should be used when operators handle aqueous solutions. Solvent and soluble protective barrier creams should be used when operators handle solvents. Sometimes emollient creams are needed to replace natural oils and fats removed from the skin.

Special work clothing

Protection of the entire body or portions of the body is often required because of exposures to extremely high or low temperatures, corrosive chemicals, and other hazards. Basic types of protective clothing are listed below.

Gloves. Many types of gloves provide protection from a variety of hazards. The selection, use, and care of gloves depend entirely upon the type and degree of hazard. Operators should not wear gloves or jewelry while working with drills, saws, grinders, or other rotating or moving machinery.

Cold weather clothing. Such clothing is to be

Figure 13-2. Retro-reflective striping is recommended for the outerwear of ramp employees.

Photo courtesy of Northwest Airlines

selected and fitted to meet applicable working conditions and temperature extremes.

High-visibility and night-hazard clothing. These garments are necessary and should be required for those employees exposed to traffic hazards (Figure 13-2).

Impervious clothing. For protection against dusts, vapors, moisture, corrosive liquids (such as battery acid), impacts, cuts, and other hazards, impervious clothing appropriate to the specific hazard must be worn. Such clothing ranges from aprons and bibs to a garment that completely encloses the body.

Safety shoes. Safety shoes should be worn by all employees where there is a reasonable probability of foot injury. Good judgment also should be exercised when selecting safety shoes. Some types will deteriorate rapidly when exposed to fire-resistant hydraulic fluids; others will not. Therefore, it is important to obtain manufacturer's specifications before buying. The sole and heel composition should be examined. Some shoes become slippery when exposed to fluids and rain. Shoes should have slip-resistant soles.

REFERENCES

American Industrial Hygiene Association. *Industrial Noise Manual*

American National Standards Institute:

Industrial Head Protection, ANSI Z89.1-1997.

Occupational and Educational Eye and Face Protection, ANSI Z87.1-1989 (R1998).

Respiratory Protection, ANSI Z88.2-1992.

National Safety Council:

Accident Prevention Manual for Business and Industry, 11th ed. 1997.

Fundamentals of Industrial Hygiene, 4th ed. 1996.

Safeguarding Concepts Illustrated. 6th ed. 1993.

U.S. Department of Health and Human Resources, National Institute for Occupational Safety and Health (NIOSH). *NIOSH-Certified Personal Protective Equipment.*

U.S. Department of Labor, Occupational Safety and Health Administration (OSHA):

Occupational Safety and Health Standards in Code of Federal Regulations, Title 29, Part 1910.

Subpart G, Occupational Health and Environmental Control (1910.24-1910.1000)

Subpart I, Personal Protective Equipment (1910.132-1910.140)

Principles and Techniques of Mechanical Guarding, Publication No.2057-1980.

WORKING GROUP MEMBERS

Contributor:

Clarence Rodrigues – Embry-Riddle Aeronautical University

AIRPORT BUILDINGS AND FIXED FACILITIES

Buildings and facilities at an airport affect the operation of the airlines and influence the attitude of the public toward air travel. Safe and attractive fixed facilities help speed the movement of passengers and the work of airline employees. In addition, pleasant, efficient surroundings can contribute to a positive attitude of the public toward air travel (Figure 14-1).

The safety and health requirements to maintain airport buildings, hangars and fixed facilities safely will be discussed in this chapter.

AIRPORT TERMINALS

The design of airport facilities is of concern to safety personnel. Proper layout and construction can help control injuries, property damage, and hindrances to operations. Ample size of facilities is desirable; otherwise, airport overcrowding can lead to incidents and injuries.

Building codes prescribe mandated construction materials, adequate fire protection, and satisfactory building egress. Before construction of a new building or transfer to an existing one, the structure should be checked to determine how well it complies with building codes. Any required changes should be made to ensure compliance with the applicable codes. The usual sources of building codes and standards in the United States are local city and state governments, federal and state regulatory agencies, the National Fire Protection Association (NFPA) and the American National Standards Institute (ANSI). (See References at the end of this chapter.)

Passenger terminals are generally laid out to provide the least possible delay to passengers from the time they arrive at the airport until they board their flights. The construction and modification of the facilities should minimize hazards that cause falls and protect people from harmful contact with machinery. Automated facilities also minimize the necessity for people to handle baggage manually.

Important safety considerations for airport terminal facilities are:

- Access for persons with disabilities. In the United States, the Americans with Disabilities Act has very specific requirements for building and safe means of access and restroom accommodation.
- Prevention of slip, trip and fall occurrences. Because of the high incidence of falls particularly involving the elderly or very young on stairways, escalators and moving walks, these means of grade change should be kept to the absolute minimum. Gradual sloping ramps are a much preferred method.
- Maintenance access is of prime importance. Safe access must be provided for heating, ventilation and air conditioning (HVAC); and all utility systems by means of platforms, overhead cranes, monorails or jib-hoists. This will permit easy and safe means for replacing heavy motors, pumps, HVAC equipment, etc. Proper access ladders and walkways must be provided for all roof-mounted equipment. Safe means of access for maintenance of lighting and windows for window washing must also be provided.
- Skylights and roof openings must be provided with protective screens.

Throughout the terminal complex, in the terminal building, and in hangars and workshops, falls suffered on stairs are one of the most common of all incidents. Falls usually are caused by undue haste, poor illumination of stairs, or defective or poorly designed steps.

Lighting in entrance areas, in stairwells, and in all other areas of a terminal building should be adequate to provide good visibility at all times. Suggested illumination levels for many operations are listed in the Illuminating Engine-ering Society of North America (IES) Lighting Handbook, Industrial Lighting, Practice for, ANSI/IES RP7-1990 (See References.)

Entrance doors should be automatic to eliminate any inconvenience to handicapped people or people with baggage. Handrails should separate the entranceway to doors. Glass doors should have an eye-catching strip across them. Doors should open away from the person going through. Exits should be treated in the same manner as entrances.

Emergency egress should conform to the Life Safety Code, NFPA 101. (See References.) Emergency exits discharging directly onto an airport ramp should be clearly identified with "emergency exit only" signs, with lettering a minimum of 2 inches high. Exits of this type should be locked from the airport ramp and must be equipped with panic hardware to allow the

Figure 14-1. Terminal buildings provide safe, attractive and efficient surroundings for passengers and airport employees.

Photo courtesy of Northwest Airlines

door to be opened in an emergency.

Escalators have been a common source of injury (Figure 14-2). These, as well as elevators and moving walks, should be installed in accordance with applicable ANSI standards. (See References) Other standards might apply outside the United States. Although the ANSI Standard for escalators permits a maximum speed of 0.64 meters per second (m/s), 125 feet per minute (fpm); a recommended precaution is to slow the escalators travel less than 0.40 m/s, 90 fpm. Other standards might apply outside the United States. This adjustment would also allow for the varying age/health of passengers and also heavy traffic usage at peak times.

Moving walkways are of special concern. Their installation may be necessary where the walking distance to gates, etc., are excessive. However, because of the high injury experience with this equipment, their use should be kept to a minimum. For existing installations, consider post-mounted emergency stop buttons at both ends. (You may consider every 50 meters.) Various "attention-grabbing" techniques to warn passengers they are approaching the end of the moving walk have been tried with varying success. Some of these devices are yellow comb plates at step-on points, yellow flashing lights at comb level by the step off, and signs with messages such as "You are approaching the end of conveyor, hold handrails." Some airports have installed handrails extending beyond the end of the moving walk at the step-off points. Distractions along the moving walkway should be minimized.

Ticket counters should be placed as close to the terminal entrance as practical to minimize the distance passengers must travel with baggage. Ticket and check-in counters should be equipped with single entrance/exits for two adjacent to workplaces avoiding the crossing of baggage belts. However, ample room for traffic on the concourse should be allowed. Ticket and baggage check-in counters should be conspicuously identified to prevent confusion that might lead to collisions and falls. Counter surfaces should be free of sharp edges or corners.

Figure 14-2. Escalators in air terminals should be easy to find and prominently marked. Escalators should travel slower than 0.45 m/s (90fpm).

Photo courtesy of Qantas

Weight scales with rounded corners and no jagged edges should be installed as close to floor level as practical, and entrances to the scales should be left free so that baggage can slide on easily. Post weigh-in baggage conveyor systems, when used, should be installed as near to the floor as possible to minimize lifting. Nonpublic conveyor installation should conform to applicable codes, and conveyors should be located so that the public does not have access to them. All pinch points on conveyors should be enclosed and the conveyors guarded from contact by personnel.

The provision of systems to maintain indoor air quality in airport facilities is of paramount importance. A system that utilizes continuous automated monitoring is the preferred approach. This must be supported by a properly qualified operations and maintenance organization.

Power services should be installed, maintained, and inspected according to the National Electrical Code, NFPA 70 (see References) or the electrical code of the local governing body. Other standards might apply outside the United States. Power panels and outlets should be guarded from access by unauthorized personnel. Power outlets on floors should be avoided; they present a tripping hazard.

Fire extinguishers and sprinkler systems should be located throughout the airport terminal building. Their placement and type should be in accordance with NFPA 10, Portable Fire Extinguishers and NFPA 13, Installation of Sprinkler Systems. (See References.) Other standards might apply outside the United States.

HANGARS, SHOPS AND FREIGHT TERMINALS

Hangars

Hangar construction is covered by federal, state and local building codes, as well as many National Fire Protection Association (NFPA) standards. Other standards might apply outside the United States.

Hangars should be designed so that aircraft can be rapidly removed in an emergency. Likewise, positioning inside the hangar should allow for the aircraft's rapid removal. Adequate aisle space; and access to fire extinguishers, hoses, sprinkler-control valves and fire-alarm stations should be maintained.

Valuable aircraft parts, engines, instruments, fixtures, and other materials should be stored in a separate building. If they are stored in the hangar, the storage area should be isolated from other departments by a physical barrier. This physical barrier should be constructed of materials that have a minimum fire rating of one hour. If the hangar is equipped with sprinklers, the storage area should be similarly equipped, but supplied by a separate system. The roof of the storage area should be watertight in order to prevent damage if the deluge equipment or sprinklers go into operation.

Storage of materials in a hangar not directly related to the aircraft or aircraft maintenance should be

Figure 14-3. Work docks and stands should offer firm footing and be protected by guardrails. Their decks should provide a firm footing or be provided with skid-resistant coverings. Operators at elevated heights must wear safety harnesses and lines when required to do so by the specific exposure or by company policy.

Photo courtesy of Northwest Airlines

avoided. Such use of hangar space, particularly during construction, has resulted in unnecessary endangerment of aircraft and the people working in the hangar.

Work docks. Work docks and stands in hangars should be constructed of fire-resistant material and should be provided with handrails; guardrails, if over ten feet in height, should be provided with toeboards (Figure 14-3). Toeboards should be provided at lower heights, beneath the open sides so persons can pass; there is moving machinery; or there is equipment with which falling material could create a hazard. The deck of these units should be made of a material that offers a firm footing or be equipped with skid-resistant coverings. Electrical service outlets should be provided on stands and docks as required, and should conform to the National Electrical Code, NFPA 70 (see References). Other standards might apply outside the United States.

Large metal complexes such as aircraft docks should be permanently bonded and grounded and secured to a permanent structure such as the hangar floor.

The hangar work dock area presents great potential for injuries due to falls. An analysis of the job site or passive fall protection systems for work performed 6 feet or more above adjacent floor or ground level OSHA CFR 1926.500.

Maintenance requirements of T-tailed aircraft have resulted in the construction of large, complex, movable tail docks. The height of the aircraft has required the installation of a lifting device or cage similar to an elevator within the superstructure of the work dock.

The appropriate classification code(s) for such hoist units can be a problem. Therefore, all applicable government rules and regulations should be consulted prior to constructing or installing one of these units. Minimum safety requirements should be established whether or not a code is applicable. Compliance codes should be adhered to at all times, without exception. In addition, the dock load capacity should be posted.

Large tail docks require extensive electrical systems for power movement of the dock and hoist, and for the outlets of maintenance equipment. The dock electrical system as well as that of the complete hangar should comply with National Electrical Code NFPA 70. (see References) Other standards might apply outside the United States.

Electrical systems. Primary sources of electrical power to hangars should be approved according to the National Electrical Code, NFPA 70, and Aircraft Hangars, NFPA 409. (See References.) Outdoor power outlets also should be code-approved with ground fault protection. All outlets should be the three-pole locking type for use with handheld lighting equipment. Hangars should have identified grounding plugs for aircraft.

Lighting. Lighting should be adequate for the task performed in each location. RP7-1983 Illumination Levels for Industry recommends 75 footcandles for incident maintenance. Extension lights should be vapor-tight and equipped with guards. Mobile lighting for use in and around aircraft should be explosion-proof.

Heating systems. The standards for heating systems and their location in hangars should be in accordance with NFPA 409. (See References.) Unit heaters approved for the area must not create a forced draft in the immediate vicinity of aircraft dockage. Heating systems should be maintained in compliance with the manufacturer's suggested standards. Coils, heat exchangers and other parts within reach of employees should be guarded.

Ventilation systems. These should be installed in accordance with NFPA 409 or equivalent, and should use power machinery suitable for the location. (See References.) Ventilation should be adequate enough to cope with climate conditions; correct ventilator installation should eliminate forced draft in work areas

Safety measures. Compressed gases such as nitrogen, carbon dioxide, acetylene, helium, oxygen and hydrogen should be stored in a designated area, and separated accordingly. Cylinders should be stored in a rack or secured firmly in place by chains. Pro-

tective valve caps should be in place. For more information see Compressed Gas Association pamphlets C-6, C-8, and P-1 (see References.)

Fire-resistant enclosures should be used for bulk storage of all flammable materials in accordance with NFPA 30 and separated from the aircraft hangar. The area should be provided with an explosion-proof electrical system, spill containment, and automatic fire protection. To protect against static buildup, all flammable containers should be provided with a grounding and bonding wire.

Fire protection, automatic and portable type, should conform to the requirements in NFPA 409 and NFPA 10. (See References)

Emergency eyebaths and showers should be provided in all areas where a person(s) may come into contact with hazardous materials.

Shops

The construction of automotive maintenance buildings should conform to the building codes required for automotive repair garages. Fire protection should conform to NFPA 10 and 13. (See References) Other standards might apply outside the United States. Automotive extra precaution is recommended, however. The ANSI standard for escalators permits a maximum speed of 0.64 m/s (meters per second) automotive hoists should meet the applicable code requirements and be provided with safety bars. Hoist capacities should be displayed prominently.

Designated areas for welding and torch cutting operations should be established and the necessary equipment provided. A hot work-permit system should be set up for performing welding/cutting operations away from designated areas. Physical separation of welding/cutting from other operations is mandatory. Adequate ventilation should be provided in welding/cutting areas; in new buildings, fixed exhaust ducts should be incorporated into design plans.

Spray painting must be performed in a designated painting room. The spray booth and spray methods should conform to the requirements of NFPA 33, Spray Application Using Flammable and Combustible Materials, and NFPA 410, Aircraft Maintenance. (See References) Paints, thinners, reducers and other flammable materials should be stored in approved cans and kept in metal lockers or wall cabinets approved for such materials.

Battery charging areas of workshops should be well-ventilated to ensure the maximum gas-air mixture that may be generated during charging is maintained below the lower explosive limits (LEL). The ventilation system should provide for automatic cutoff of charging equipment in case the ventilation blower or fan fails. Such battery charging operations should be located in a separate building or in an area (as described in NEC 513-2d) that is ventilated adequately and effectively cut off from the hangar, shelter or nose dock by walls or partitions.

Freight terminals

A great deal of airfreight is handled in old buildings that were intended for other uses. Freight is usually handled manually, and with lift trucks and carts. Power conveyors and automatic handling systems are an acceptable standard. Efficiency is maintained by good supervision and tight scheduling, along with excellent housekeeping.

Satisfactory airfreight terminals can be constructed similar to a commercial warehouse or motor freight terminal. Construction should comply with the applicable building codes and NFPA standards.

Buildings should incorporate a layout that will expedite the handling of both inbound and outbound freight. Power conveyors should be utilized where possible. The system should be designed to eliminate multiple handling of freight.

Aisles and separations in airfreight terminals should be sufficiently wide enough to help contain a fire within a small area. Dividing racks into smaller segments by means of fire partitions is another method of limiting a fire to a small area. Stored material should be kept at a height where they can be accessed by automatic or manual fire-extinguishing systems. Using noncombustible materials for racks, shelving, and pallets will reduce the fuel that could feed a fire and thus increase the possibility that stored material could be removed in an emergency.

Sprinkler protection can be effective in controlling ceiling temperatures and preserving structural integrity, if selected on the basis of hazard classification and the vulnerability of the roof or ceiling construction. (High-expansion-type forms are finding increased acceptance in warehouses and airfreight terminals.)

The type of fire protection system required is dependent upon the type of material stored, the height and density of stored materials, aisle width, and construction material of racks and pallets, etc. Rack Storage of Materials, NFPA 231C, applies to a broad range of combustibles stored higher than 3.6 m (12 ft) on racks. Storage on plastic pallets or plastic shelves, bin storage, or shelf storage is not covered; nor is storage of high hazard materials, such as tires, roll paper or flammable liquids. For storage on racks shorter than than 3.6 m, see Installation of Sprinkler Systems, NFPA 13 (See References and Chapter 6, "Fire Prevention and Protection")

Standard docks should be provided for both inbound and outbound freight. Spring-loaded dock plates should be built-in so that material-handling equipment can move in and out of trailers and trucks with minimum restrictions. If standard dock plates are used, they should have cleats or pins that ensure proper locking into position.

The ideal facility provides a dock that permits transfer of freight to and from truck beds at the same level as the terminal floor. Arrangements should be provided for station wagon, van and pickup truck heights as well as for higher wheels-under-bed trailers. Truck wells or dock levelers can fill this function.

Forklifts or similar removal equipment must not enter a docked vehicle unless the wheels on the vehicle are chocked to prevent movement.

The area in front of the dock should be kept clean and free of any broken boxes or cartons. Adequate lighting is required to enable the truckers to back into a position without hitting adjacent trucks or people.

Overhead doors in the freight terminal should be counterbalanced to facilitate opening or inadvertent closing.

Adequate ventilation should be provided to dispel fumes and gases generated by combustion-engine equipment moving in and out of the facility. Consideration should be given to the use of electrically operated material-handling equipment in a freight terminal, rather than engine-driven equipment.

Stacking pallets of freight for cargo liners requires a hoist or platform to build up the pallet stack. The hoists, though excluded from the ANSI standard for elevators, A17.1 (see References), do come under local or state codes in some instances. Before the hoist or platform is installed, the appropriate code should be checked and any compliance requirements met. The main issue surrounding pallet buildup hoists is the requirement to have a skirt guard fully enclose the unit when it is being raised or lowered.

Smoking should be prohibited in the general air-freight building. Specific areas should be designated where smoking is permitted.

For information on procedures for handling freight in dock areas, see Chapter 19, "Air Cargo Operations."

RAMP SERVICE FACILITIES

Aircraft service facilities located in ramp pits are becoming common both at existing airports and at new facilities. These pits do much to improve the efficiency of the general ramp operation by eliminating the need for mobile ground equipment.

The pits should be located as near to the aircraft service connections as possible. Proper planning will eliminate the necessity for cables or hoses to be lying on a ramp (Figure 14-4).

Periodic checks of the pits should be established in order to determine and control possible fuel or fume accumulation.

Electrical power pits in ramps should have a stationary motor-generator. Mobile ground power units should not be permitted. The motor-generator should be located alongside the terminal concourse. Maximum voltage is 220. Warning signs must be posted per National Electrical Code 1999, section 110-17(c). Electrical wiring should conform to National Electrical Code, NFPA 70, for Class I, Group D, Division I hazardous locations. (See References.)

Fuel pits should be located a minimum of 15 m (50 ft) away from a terminal building or concourse (Figure 14-5).

The flow of more than one fuel through a single line is prohibited. Separate pumps, single lines and standard couplings will be utilized for each type of fuel. Color codes should be mandatory and should comply with accepted standards.

The compressor, air conditioning and heating units for the ramp pits should be located in the concourse structure. Their installation should comply with applicable codes.

The connector on the air starter should be designed to eliminate the possibility of an accidental disconnect during start-up.

CONSTRUCTION SAFETY

Contractors engaged in new construction, major alterations or demolition of buildings should be required by contract to operate in a safe and controlled manner. Requirements should be based on established work practices, the Associated General Contractors of America Manual on Incident Prevention in Construction, and NFPA 241, Safeguarding Construction, and Demolition Operations (see References).

Contractors should be advised of the dangers associated with prop and jet blast. They should be given assistance in developing a program for materials handling, general cleanup and housekeeping, and fire protection.

Before any construction commences, an inspection of the site should be conducted. All hazards should be identified, and steps should be taken to protect employees and the general public from them. Providing barricades and overhead protection should be the responsibility of the contractor. Constant monitoring of construction areas should also be provided to ensure safety standards are being met.

Construction offices and storage sheds should be of noncombustible materials. Heaters should be of the approved type only, and installations should comply with applicable codes. L-P gas interior heater and temporary outside heaters should comply with NFPA 58 Liquefied Petroleum Gases. (See References)

All contractor and subcontractor employees should be required to operate within the terms of the contract. Construction sites located on ramp areas present potential foreign object debris (FOD) dangers to jet aircraft. FOD control by the contractor is essential.

Temporary structures

Temporary structures anywhere at the airport should be of sound, substantial construction. Illumination, heating and fire protection should conform to applicable standards. Temporary structures, for the most part, are of materials with low fire-resistive ratings. As a result of their potential fire hazards, they should not be constructed near facilities housing high-value merchandise, aircraft or important operations.

Temporary structures should be removed as soon as they are no longer needed.

MAINTENANCE OF OPERATIONAL FACILITIES

The airport authority should establish an airport-wide system for the investigation and reporting of all incidents – those involving members of the general public, contractors or employees. This will enable the authority to analyze the types and trends of incident and equipment involved. These reports should be reviewed by airport management and safety personnel so that any conditions such as unsafe vehicle operations, improper materials handling or tripping hazards will be promptly corrected.

Figure 14-4. Covers for service facility pits should be of lightweight metal and should require no more than 130 newtons (30 lb) to lift.

Photo courtesy of Northwest Airlines

Figure 14-5. Fuel pits should be located at least 15 m (50 ft) away from the terminal building or concourse.

Photo courtesy of Northwest Airlines

All buildings are subject to deterioration; therefore, procedures should be established so that any evidence of structural failure or deterioration is reported at once and repairs made immediately. Safety and maintenance personnel should conduct regular surveys aimed at identifying unsafe conditions, hazard detection, hazard analysis and hazard control (i.e., preventive maintenance programs should be developed and implemented).

Building maintenance

Roofs should be inspected annually, as well as after violent storms. It is important to look for signs of excessive movement, strain, or overloading in the structural members of roofs, framing, trusses and floor joists.

All building doors should open outward. Exits should be well-lighted and marked. Illuminated exit signs should be installed for all emergency exits and passageways, as required by NFPA and local building codes. Panic hardware should be installed on the inside of exterior doors used for emergency exits. Exit doors should be kept unobstructed and unlocked in the direction of egress. Doors should not open immediately onto a flight of stairs, but on a landing at least the width of the door. Emergency evacuation plans should be posted in all buildings. Particular attention should be paid to the correct design and orientation of these evacuation plans (i.e., consideration of human factor type errors that could result).

Windows designed to be opened should be kept in good operating condition to prevent sticking. Broken windows and screens should be replaced and repaired as soon as possible.

Bolts used for attaching window washers' belts should be fastened to reinforced window frames – not to exterior walls. These bolts should be installed on all windows above the ground floor unless the windows can be safely cleaned from the inside without the window washer having to lean out. When working above the ground floors, window washers should wear safety harnesses. All harnesses should be checked periodically for defects, and should be stored and maintained in good condition.

A sufficient number of approved fire extinguishers should be located in all buildings. They should be easy to reach in case of fire. A clear access passage of at least 0.9 m (3 ft) wide should be maintained for them at all times. Extinguishers should be serviced according to NFPA codes. Emergency lighting should be located at all egress points.

Stairs, ladders and ramps

To reduce the possibility of incidents, stairs and steps should be kept clean and free of obstacles or slippery substances at all times. Stair surfaces should be kept in good repair. Loose boards, unsecured treads, protruding nails, and torn or worn stair treads should be repaired or replaced immediately upon detection.

The use of open risers should be avoided whenever possible. Slippery treads should either be replaced or coated with slip-resistant surfacing materials. Stair nosings should be securely fastened, and rounded or beveled to prevent people from catching their heels on the treads.

Where there is less than 2.1 m (7 ft) of headroom over stairs, the obstruction should be padded, suitable warning signs posted and the overhead obstacle clearly painted in contrasting colors.

Both interior and exterior stairways should be sufficiently lighted so that all treads and landings are clearly visible. For indoor stairs lights, there should be three-way switches at both the top and the bottom landings.

Floors

Cracks, holes, bumps and other defects in floors can endanger safe handling of materials and cause injury to employees. Therefore, floors should be kept in good repair. To reduce the chance of falls, floors should be cleaned with slip-resistant substances or covered with slip-resistant materials.

Misused floors will wear excessively. Here are some misuses or abuses that should be avoided:

• Use of heavy equipment or storage of heavy materials may overstress or weaken floors.

- Spilling acid or oil on nonresistant floors can damage their surface.
- Dropping heavy or sharp articles, such as towbars, can cause floors to spall or crack.

The maximum allowable loading per square foot of floors should be prominently displayed in all storage areas. In cases of unusually heavy, concentrated loads such as safes, a structural engineer should analyze the condition and determine if floors are strong enough. The engineer should recommend means of additional support, if necessary, or alternate storage areas. Recommended load limits never should be exceeded, and heavy loads should be spread over large areas instead of being concentrated in one area. The weights of hoists and other equipment suspended under the floor should be considered in determining the floor's safe load limit.

Joists, beams or subflooring in old structures may have deteriorated to such an extent that the original maximum load limits are no longer valid. Where evidence of deterioration exists, a structural engineer should determine the safe load limit or the need for repairs.

Floor openings should be guarded by railings and toeboards unless they are fully protected by other means. After dark, floor openings should be illuminated adequately to prevent anyone from falling through. Any telephone and electric conduit inspection plates should be kept in safe condition to prevent tripping.

Floors and stairways should not be waxed if this causes a slipping hazard. Appropriate slip-resistant waxes or other finishing compounds should be used. The primary reasons for waxed floors becoming slippery are too much wax, excessive polishing of waxed floors and accumulations of liquids or other slippery substances.

MAINTENANCE OF EXTERIORS AND GROUNDS

Exteriors

Metal chimneys or stacks should be checked frequently for corrosion and other defects that could possibly cause collapse. Steel guy wires, used to brace stacks, also should be checked periodically for defects.

Wire braces and guys should be properly placed and kept under proper tension. In areas where vehicles will pass under guys, the wires should be kept at least 3.5 m (12 ft) above the ground by means of posts. This gives sufficient clearance to prevent vehicles from catching the guys and toppling the chimney.

All chimneys and stacks should be fitted with spark arrestors if located in areas where flying sparks may be a fire hazard. Adequate lightning protection systems, connected to low-resistance grounds, should be provided on all high objects.

Chimneys, water towers and other high obstructions should be painted in a proper color-code as required by the authority having jurisdiction. They should be equipped with appropriate lights for after-dark identification.

Grounds

The grounds, particularly around buildings and flammable or explosive storage areas, should be well guarded and kept free of nonessential flammable materials.

All construction work on grounds should be clearly identified by signs. Construction areas should be protected by barriers suitably marked with reflective materials or illuminated – or both – for easy visibility after dark.

Roads and walkways should be kept clear and in good repair. They should be defined, and adequately lighted at night to ensure pedestrian and vehicle safety.

During winter months, snow and ice removal equipment and materials should be kept immediately available. Sand, cinders, ashes, or sodium or calcium chloride should be used to cover icy roads and other slippery surfaces. Icicles should be removed from above doorways. Persons performing this task should protect themselves against injury from icicles and other falling debris.

Lines, wires and similar obstructions strung less than 2.1 m (7 ft) above the ground should be clearly marked. Normally, low-hanging wires should not be located over roads or sidewalks, but where necessary, particular care should be taken to avoid contact by passing vehicles or people. Electrical lines must be either elevated enough to clear traffic or buried underground.

REFERENCES

American National Standards Institute, 1430 Broadway, New York, NY 10018

Elevators, Escalators, and Safety Code for, ANSI A17.1-1996.

Marking Physical Hazards, Safety Color Code for, ANSI Z535.1-1998.

Industrial Lighting, Practice for, ANSI/IES RP7-1990.

Consumer Turf Care Equipment- Walk-Behind Mowers and Ride on Machines with Mowers ANSI/OPEI B71.1-1998

Associated General Contractors of America, Inc., 1957 E St., NW, Washington, D.C. 20006.

Manual of Accident Prevention in Construction, 8th ed., 1999.

Compressed Gas Association, 1235 Jefferson Davis Hwy., Arlington, VA 22202

Requalification of DOT-3HT Semless Steel Cylinder, CGA.C-8-1999

Safe Handling of Compressed Gas in Containers, CGA - P-1, 1999

Visual Inspection of Steel Compressed Gas Cylinders, CGA C-6, 1999

"Documented Personnel Exposures from Airport Baggage Inspection Systems." October 1976. Publication #77-105. Available from National Technical Information Service, Springfield, VA 22161; Stock No PB 274-202.

Illuminating Engineering Society of North America, 120 Wall Street, New York, NY 10005-4001.

IES Lighting Handbook

National Fire Protection Association, Batterymarch Park, Quincy, MA 02269

Aircraft Fuel Servicing, NFPA 407, 1997.

Aircraft Fueling Ramp Drainage, NFPA 415, 1997.

Aircraft Hangars, NFPA 409, 1995.

Aircraft Maintenance, NFPA 410, 1997.

Aircraft Rescue and Fire Fighting Vehicles, NFPA 414, 1997.

Evaluating Foam Fire Fighting Equipment on Aircraft Rescue and Fire Fighting Vehicles, NFPA 412, 1997.

Installation of Sprinkler Systems, NFPA 13, 1996.

Life Safety Code, NFPA 101, 1997.

Liquified Petroleum Gases, Storage and Handling of, NFPA 58, 1995

National Electrical Code, NFPA 70

Portable Fire Extinguishers, NFPA 10, 1994

Rack Storage of Materials, NFPA 231C, 1995

Safe Guarding Construction and Demolition Operations, NFPA 241, 1996

Spray Application Using Flammable or Combustible Materials, NFPA 33, 1995

U.S. Department of Health and Human Resources, National Institute for Occupational Safety and Health (NIOSH), Cincinnati, OH 45226

U.S. Department of Labor, Occupational Safety and Health Administration (OSHA) 200 Constitution Avenue, Washington D.C. 20210

A passenger terminal may be as simple and unpretentious as a room next to the hangar, or a hangar annex used by the fixed-based operator as a gathering point for clients and customers. In a small community, a local service terminal building of modest size with adjacent ramp area may be used to process and board passengers. However, the majority of air travelers pass through the extensive building and ramp complexes of the major air terminal.

The material in this chapter can be applied in both small and large terminals. Either hold tremendous potential for loss of life and/or litigation if systems fail or a major disaster occurs.

Titles II and III of the Americans with Disabilities Act of 1990 (ADA) specify that people with disabilities must be given an equal opportunity to benefit from all available services, programs, or activities. This includes transportation and access to public areas. For more information, see the *Guide to Disability Rights Laws*, U.S. Department of Justice, Civil Rights Division, Disability Rights Section, May 1997, Washington.

PUBLIC AREAS

To prevent injuries to the public in terminals, the following basic considerations apply:

- No hazards should impede the flow of passengers from the street to the aircraft. This is important because most injuries are caused by falls.

- Facilities should be clean and well maintained to eliminate hazards that deteriorating facilities or poor housekeeping can cause.

- Efficient handling of passengers is essential. Otherwise, crowding will pose an incident hazard.

- The fact that people of all ages – from infants to the elderly – use air terminals must be considered in their design. Many people are new to air travel, and others may be confused when facilities at a strange terminal differ from those with which they are familiar. Also, airport sightseers, of which there may be fairly large numbers, need protection from personal injury (Figure 15-1).

- In designing the airport, remember that rapid travel upsets people's routines and often causes fatigue. Therefore, terminals should be built to allow for reduced awareness on the part of the users. For example, consideration should be given to providing more footcandles of illumination than comparable areas in other kinds of facilities would require. The decor of air terminals should incorporate as much bright, distinctive color as possible, particularly on stairs, at exits and in corridors. Signage should incorporate human factors-based symbolism, be noticeable at a glance and easily read (Figure 15-2).

Figure 15-1. Crowds of sightseers and well-wishers, as well as passengers, must be safeguarded from injury in airport terminals.

Photo courtesy of Northwest Airlines

Parking and vehicle-access areas

Public parking areas around airport terminal buildings should be situated as close to the terminal entrances and exits as is consistent with the general plan. Parking lots should be well surfaced, drained, and lighted. They should have pedestrian walkways and car parking lanes that are conspicuously marked and maintained (Figure 15-3). Parking areas should be kept free of debris. Snow and ice should be cleared as they build up during a storm. Parking areas should be patrolled to detect unsafe conditions, and to discourage vandalism and theft.

Where multi-story parking facilities are needed, care should be taken to ensure that the direction of traffic flow is obvious. Adequate illumination must be provided in multi-story parking lots; it is possible that interfloor illumination will be required 24 hours a day.

Traffic patterns in and near parking areas should be circular. The entrances and the exits should be on opposite sides of the lot. Traffic should flow in accordance with designed patterns to avoid collision possibility and pedestrian injury. Pedestrian exits from parking lots and walkways to terminal entrances and exits should be controlled by traffic signals during peak hours of traffic.

For vehicles approaching the terminals, dual ramps – one for arrivals and one for departures – help pre-vent incidents. The traffic on the ramps should be circular.

Cab ranks should be maintained as near to the exit of the terminal as is compatible with general congestion in the area. It may be necessary to have a holding area at some distance from the terminal building and that cabs be called by traffic marshals.

Passenger-access areas

It is essential that no obstacles are placed in the path of passengers entering the terminal building. Porter (skycap) locations, benches, dolly carts, dispatcher locations and other obstacles must be kept clear of terminal entrances. There should be unimpeded access to the terminal entrance at all times.

Pads or mats for automatic door-opening devices should be well maintained to reduce the possibility of tripping. Most floors become slippery when wet. Therefore, runners or floor mats may be advisable during inclement weather. Floors should be finished with slip-resistant floor wax, as tested and classified by Underwriters Laboratories (UL) or similar agencies.

Corridors, stairs, escalators and moving sidewalks leading from the ticketing area to the aircraft boarding area should be kept as clear and orderly as possible. Baggage dollies, wheelchairs, lounges, ash-

Figure 15-2. Signs that guide passengers should be prominently displayed.

Photo courtesy of United Air Lines

Figure 15-3. Circular traffic patterns produce a regular traffic flow and help prevent accidents.

Photo courtesy of Northwest Airlines

trays, refuse containers, planters and other furniture should be positioned well clear of the main thoroughfares (Figure 15-4).

Smoking/non-smoking areas must be clearly designated.

An adequate number of ashtrays and refuse containers should be placed throughout the passenger area. Rubbish containers should be identified with "for trash only" signs. Ashtrays should incorporate an extinguishing feature; they should be fastened to a substantial base that will prevent tipping if they are accidentally bumped. Self-closing cigarette containers or standard sand buckets are recommended. These conveniences must be kept in good condition.

Stairways should be covered with slip-resistant material, and handrails must be maintained in good condition. Escalators should be maintained according to the manufacturer's standards. They should be suitably guarded and posted with signs indicating the location of emergency stop buttons and the correct method of using them. Similar precautions apply to moving sidewalks. Load limits should be posted on elevators and escalators. Wheeled vehicles, such as baby carriages and wheelchairs, should be prohibited from escalators. (See ANSI/ASME A17.1-1996. See References.)

Escalators may pose a hazard to young children and senior citizens; special arrangements should be made for these groups. Signs may redirect them and the people accompanying them to elevators. In larger terminals, passenger service carts may be made available from the traffic staff department; signs in the terminal should indicate the availability of this service.

Lighting in all areas of a terminal building should be well maintained. An emergency lighting system should be installed as a precaution if a primary power failure occurs.

Warning signs and barricades should be placed around areas where remodeling or other construction work is taking place.

Figure 15-4. Traffic areas should be kept clear and orderly. Floors should be finished with slip-resistant preparations and kept free of all impediments to passenger movement.

Photo courtesy of United Air Lines

Baggage-handling areas

Baggage weigh-in facilities should minimize the need for passengers and employees to lift and handle bags (Figure 15-5). Where fully automated combined check/weigh conveyors are used, consideration should be given to locating the baggage-height gauge at the front of the check-in desk. A split curtain or screen should be installed to discourage youngsters from climbing onto the mechanized handling system.

In baggage claim areas, the carousels, conveyors and dispensers should have a pre-start warming horn, signs warning of the hazards, emergency shutoffs, be inaccessible to children, and have built-in guards that prevent hazards to passengers or operators.

Baggage claim areas should be patrolled by company personnel. If a "hang-up" occurs to baggage, passengers should not be allowed to scramble over a moving carousel or other conveyor to retrieve baggage that has stopped moving. Power equipment for the baggage conveyor systems should be positioned and guarded to minimize passenger hazards.

Arriving passengers often are accompanied by relatives and other greeters, so the baggage pickup area can become quite congested. Security officers may be utilized to verify bag ownership and encourage safe passenger flow. Easy access to and from the bag-

Figure 15-5. Weigh-in facilities located at check-in counters reduce the need for passengers to handle and lift baggage. Guards and emergency shut-off prevent any hazards to workers.

Photo courtesy of Delta

gage claim areas must strike a balance with security considerations.

Exit doors from the baggage pickup area should be automatic, and rail guards should lead to the doors. Canopied areas should be well-lighted and protected. Stairs to the street level should be slip-resistant and provided with handrails. Direction signs should be illuminated at all times, and pedestrian walkways should be indicated, outlined and controlled. Luminous signs that remain visible in the dark may be advantageous if a power failure occurs.

Public service areas

Terminal buildings are designed to facilitate the movement of passengers on and off aircraft, speeding them on their way but also catering to their needs in the process. Extensive service areas are a necessary part of this process.

Service areas usually include restaurants and bars, observation decks, lounges, rest rooms, nurseries, workshops and kiosks. The public areas around them should be well-maintained.

Standard railings should be considered for protect-

ing glassed areas from damage by carts, baggage or crowds. It is important that glass doors and walls be noticeable so that pedestrians do not incidentally collide with them.

Where national or local government regulations apply to public areas, the owner, landlord, tenant or concessionaire should ensure such regulations are enforced. These areas include restaurants, cocktail lounges, kitchens, passenger elevators and escalators. Regulations often require controlling maximum seating capacity, providing an adequate number of exits and free egress from those exits, and marking aisleways, routing and electrical wiring maintenance. Often regulations call for regular inspection or certification.

Lounges, restrooms and nurseries should be kept clean, have slip-resistant floors, and be furnished with ample supplies.

Ideally, lounges should be located away from crowded corridors and ticketing areas, because their prime function is to provide a resting area for passengers. However, restrooms should be in high-traffic areas and well lighted to discourage vandalism and harassment of users. Nurseries should be in well-controlled, high-security areas.

OPERATING AREAS

Airport personnel should be aware of special safety precautions to be followed when guiding passengers to and from aircraft. Weather conditions also may require special safeguarding practices.

Gate positions and holding areas

Potential incident situations at gate positions must be identified and avoided. For example, once a flight has been announced, passengers may crowd at the ramp entrance. Procedures should be established to reduce such hazards, and employees should exercise judgment in handling passengers.

Signs designating gate areas should be prominent. In order to provide adequate control and accommodation in holding rooms, occupancy should be limited to passengers only if feasible.

Signs designating "No smoking beyond this point" should be located at doors leading to ramps or to aircraft loading walkways and supplemented by pictograph symbols and broadcast messages in languages appropriate to the terminal.

When practicable, there should be no steps from the holding room to the tarmac, but rather a gradual incline. Furnishings within the holding area should be solid, but positioned out of the traffic path leading from the plane. Floor coverings should be slip-resistant, even when wet. Carpeting is one solution.

Aircraft loading ramps

Aircraft loading walkway construction should be in accordance with local legislation or authority. As an example, NFPA 415 (see References) applies in the United States. Other standards might apply outside the United States.

Because of the large number of moving vehicles on them, ramps or aprons are potential sources of passenger injury. For this reason, it is essential to position equipment around aircraft so that passenger walkways are completely unobstructed. Walkways should be clearly marked with movable tapes or otherwise defined to keep passengers clear of the aircraft and support equipment. This is particularly important with propeller-engine or rotary-winged aircraft. Walkways should not lead over tripping hazards or low-lying portions of the ramp where puddles may have accumulated.

Passenger pathways on the tarmac should not pass under aircraft wings where engines and other pendants might cause head injuries. Wheel chocks, fire extinguishers, and other obstacles should be kept well clear of pedestrian pathways. Service vehicles should not cross tarmac-level boarding and disembarking routes. Passengers must be protected against jet engine blast (Figure 15-6).

Equipment not in use at an aircraft should be parked in its designated parking space. It should be out of the way of passenger and vehicular traffic.

Ramp areas adjacent to parked aircraft should be well lighted by glare-resistant lamps to allow disembarking passengers to identify the appropriate entrances. Such lamps should be positioned as high as is practical. Lighting provided by floodlights mounted on the terminal building would be preferable to avoid obstruction hazards presented by pole- or standard-mounted lights, keeping in mind that the lights must be positioned downward to prevent light pollution in the flightdecks of arriving and departing aircraft.

Airport ergonomic design considerations

A long lasting concern of airlines is the interface between baggage handling personnel and the equipment they use in performing their duties. Good equipment and good procedures can be instrumental in reducing injuries and improving efficiency of those who handle baggage. Individually, airlines have observed various standards for baggage systems in order to improve ergonomic characteristics of the equipment used. As a result, work-related injuries have decreased, along with concomitant decreases in costs to employees and employers, while productivity has increased.

Figure 15-6. Loading bridges protect passengers from the environment, aircraft and operational equipment.

Photo courtesy of Delta

The experience gained by airlines has not automatically transferred to other industry parties who are involved in the design and construction of new airport facilities where baggage handling takes place. Listed below are areas that require ergonomic design considerations or remodeling.

- Curbside conveyor systems
- Check-in counters conventional and kiosk style
- Check-in counters feeder belt conveyor systems
- Load conveyors
- Carousel conveyors
- Incline plate make-up device
- Sortation laterals
- Inbound oversize slide
- Bag-room floor

Mechanical aids as an assist to ergonomic design are in a constant state of development. At present, flexible conveyors used in the aircraft hold are on the market. Mechanical arms for lifting bags in the baggage make-up room are at the trial stage, as are mechanical aids for lift on/lift off of wheelchair passengers. It is important for designers, manufacturers and users to take developing technology into consideration at the design stage.

Weather factors

Roadways, walkways, and exitways should be protected from the elements, insofar as is practical.

Ramps, parking lots and areas of vehicular congestion should be constructed and surfaced to drain rapidly. They should be kept free of any obstruction to the free flow of traffic or to snow removal equipment. Ice and snow removal should continue as a storm progresses, so that no areas become blocked. During snowstorms, ramps should be cleared and icy patches sanded or chemically treated before passengers are allowed to pass to or from an aircraft. If responsibility for snow and ice clearance rests with an airport or a local government authority, conveniently located containers of salt or sand mix might be provided for staff use during weekends or nights, when it often is difficult to alert the responsible authorities.

During periods of high winds, it is advisable to park aircraft on the windward side of the terminal finger. This helps to ensure passengers and operators are not injured by gravel blown off the roofs. (In extreme cases, part of the roof might peel off and cause serious injury.) In addition, parking aircraft upwind of the holding room will have the added advantage of protecting embarking passengers.

Operations should be suspended while an electrical storm is in the immediate area of a ramp. High projections – preferably metal structures designed to conduct electrical discharge to the ground – can help reduce personal injury, equipment damage and fire exposure. Electronic equipment is available that measures the probability of lightning strikes within various distances of the terminal.

MEDICAL FACILITIES

The extent to which medical or first aid services will be provided depends on terminal size, passenger traffic, and resources available outside the airport.

Smaller airfields must make arrangements with a local medical authority for support in case of serious incident or illness. In larger airports, permanent facilities, staffed with fully qualified medical personnel, might be provided. Services to passengers can include not only first aid and emergency medical treatment, but also – in the case of international passengers – inoculation and issuance of medical certificates.

In most major domestic and international airports with high traffic frequency, a comprehensive medical center would be an integral feature of the airport plan to provide an immediate response to an airfield crash or inbound aircraft with medical emergencies on board.

Airports served by airlines must have a medical plan to deal with emergencies involving the largest aircraft that might land there. In the United States, the Federal Aviation Administration requires plans that detail the capabilities of medical facilities, rescue units, available safety/security/military personnel and nearby shelter areas.

Every airline must also develop methods "appropriate to its own operations and aircraft" for transporting physically handicapped people, as well as evacuating them from the aircraft in the event of an emergency.

SECURITY

Airports and airline operators have been a target of terrorists and other individuals. Aircraft have been hijacked and destroyed, passengers held hostage and killed, both in aircraft and in airport lounges.

Security measures

Measures adopted to counter threats to passenger safety have varied widely. Some are publicized; others, for security reasons, kept secret. Protection includes:

- Laminated bullet-resistant glass for check-in points
- Treated-surface glass, which reduces flying glass following a bomb blast
- Closed-circuit TV to monitor movement in critical areas
- Physical and electronic searches of passengers, hand baggage and checked baggage.

NOTE: COMBINED MAGNETIC FIELD DOORWAYS, HAND-HELD DETECTORS, TV APPARATUS AND VAPOR DETECTORS HAVE BEEN DEVELOPED TO ASSIST IN SEARCH PROCESSES. WITH SUCH EQUIPMENT, STAFF OPERATORS MUST BE PROVIDED WITH THE USUAL FILM BADGES TO MEASURE THEIR EXPOSURE TO X-RAY RADIATION.

There are both financial and social constraints on the degree of security possible. For example, it may be unacceptable to restrict access to airfields to ticket-holding passengers only.

The extent to which a terminal operator should develop security measures is dictated by the Transportation Security Administration (TSA) or other agencies. Specialist advice can be obtained from local and federal law enforcement agencies and/or consultants.

It is essential that all airport personnel be familiar with procedures for evacuating terminals, even when it is difficult to hold actual rehearsals because of the presence of passengers. Periodic training exercises should be conducted to ensure all staff are adequately skilled in this area.

REFERENCES

American National Standards Institute:
  Industrial Lighting, Practice for, ANSI/IES RP7-1990.
  Elevators, Escalators, and Safety Code for, ANSI/ASME A17.1-1996.
National Fire Protection Association:
  Airport Terminal Buildings, Fueling Ramp Drainage, and Loading Walkways, NFPA 415, 1997.
  Aircraft Hangars, NFPA 409, 1995
  Installation of Sprinkler Systems, NFPA 101, 1997
  Life Safety Code, NFPA 101, 1997
  Liquefied Petroleum Gases, Storage and Handling of, NFPA 58, 1995
  National Electrical Code, NFPA 70
  Portable Fire Extinguishers, NFPA 10, 1994
  Rack Storage of Materials, NFPA 231C, 1995
  Safe Guarding Construction and Demolition Operations, NFPA 241, 1996
  Spray Application Using Flammable or Combustible Materials, NFPA 33, 1995
  Aircraft Maintenance, NFPA 410, 1994
Illuminating Engineering Society of North America. *IES Lighting Handbook*
National Technical Information Service, *Documented Personnel Exposures from Airport Baggage Inspection Systems.* 1976. Publication #77-105.
U.S. Department of Health and Human Resources, National Institute for Occupational Safety and Health
U.S. Occupational Safety and Health Administration (OSHA)

RAMP OPERATIONS
AT PASSENGER TERMINALS

CHAPTER 16

Parking and moving aircraft
Engine starting
Foreign object debris (FOD)
Prevention techniques
Aircraft loading and unloading
Passenger loading
Employee procedures at ramps
Cargo handling
Weather
Ground equipment movement
Organizing vehicular traffic
Operating requirements
Driving rules
Aircraft ground handling
Headset operator (aircraft pushback)
Hand signals
Visibility
Procedures
Training
Parking
Servicing aircraft
Fueling operations
Passenger water and lavatory services
Deicing operations
Communications in ramp areas
Protection against jet blast and noise
Reducing jet blasts
References

Whether hundreds of passengers and tons of cargo or only one passenger and his or her luggage are to be loaded, the principles are the same for safe operation at passenger terminal ramps. This chapter is intended to help minimize incident exposure to passengers and employees; and damage exposure to aircraft, ground support equipment and facilities in the ramp areas of passenger terminals.

During ramp operations, various hazards are present due to the nature of the work and the equipment and tools involved. Other factors involve weather conditions, day and night operations, schedule pressures, and the various aircraft systems. Operators are exposed to, noise, vibration, jet blast/prop wash and hazardous materials. Potential hazardous energy sources such as electricity, hydraulic pressure lines, compressed gas, rotating components, chemicals and vehicle operations must be identified and controlled. Improper lifting procedures while handling baggage,

tires, brakes and other aircraft parts can cause sprains and strains to the body. The potential for falls increases while performing work at elevated levels around aircraft. Additionally, human factors, both physical and mental, may also affect ramp operations.

Parking and Moving Aircraft

Factors that determine parking position of aircraft are:

- Position of fueling or other fixed services
- Position of the jet bridge for passenger disembarking
- Positions of passenger gates and boarding and disembarking paths
- Position designated for adjacent parked aircraft
- Access and working positions for aircraft service, passenger service, and cargo-handling vehicles and equipment

One variable factor that affects the parking position of aircraft is the wind. Both the direction and the velocity of the wind should be taken into account.

At times, such factors as heavier-than-normal ramp traffic, repairs on ramps or terminals, construction operations, or snow and ice accumulation make it necessary to park aircraft in less than desirable areas. In such instances, special care and alertness are required of ramp crews. Some precautionary measures to use are temporary lighting, roped or taped walkways, and additional staff to direct passengers around or clear of the hazardous areas.

The principle hazards to be guarded against while moving aircraft on ramps are engine ingestion, collision and prop wash or jet blast. Whether aircraft are taxiing or being towed, their wing tip, tail and nose positions should be closely watched so that warning can be given against impending collision with other aircraft, fixed installations, or ground equipment. (See "Communications" section near the end of this chapter for further details.)

When aircraft are taxiing, prop wash or jet blast hazards are to be guarded against. Wash or blast can slam doors closed, knock people down, and cause turbo-prop propellers to rotate. The strong turbulence also can cause unsecured equipment to roll or blow over or cause foreign object debris (FOD) to be blown into the engines of other running aircraft or onto people in the area. Such occurrences can result in injured people and damaged equipment. Prevention requires well-planned standard operating procedures executed by trained and disciplined ramp crews.

Engine starting

During engine start at a ramp, alertness on the part of ground and flight crews and precise operation are required to ensure the safety of passengers, employees and equipment.

At all times, the engine-start sequence should be supervised by a capable and reliable person. It is the duty of this person to make sure all personnel are clear of danger areas. Further, all unnecessary ground support equipment is to be removed from the vicinity of the aircraft. The person in charge must maintain close watch on the aircraft and its vicinity to make sure no one enters the danger areas once the start sequence has commenced. This person also must watch for boarding or disembarking passengers at other aircraft in the vicinity, particularly those behind the one being started. Jet blast or prop wash could injure passengers or employees in the area.

Before the actual engine-start sequence begins, the person controlling the start procedure must ascertain all of the following are properly secured: passenger entrances, galley service doors, cargo compartment doors, and access and service panels.

A common method of communicating with the flight captain is hand signaling and voice. It is essential that standard authorized signalling procedures be followed (Figure 16-1). The flight deck crew and ground employees controlling the start sequence should have a firm knowledge of the signals, particularly those used to indicate the presence of an emergency or a hazardous condition. Signals must be executed in a clear, sharp manner that precludes confusion between signals. Ramp operators must avoid lazy half-motions. They

also should refrain from making any motions such as adjusting their head cover or rubbing their nose or face, because such gestures may be misconstrued as a signal. (Interphones for direct contact between the ground and the flight crew may also be used.)

Communications between an aircraft and ground often require telephone or radio communication sets. It is a prime requisite that such sets be maintained in good working order, and that all personnel responsible for their use understand both their operation and the procedures laid down for their use. Unnecessary conversation during the start sequence must be avoided since it could cause misunderstanding of important commands or acknowledgments. (See the "Communications" section later in this chapter.)

Operators charged with the responsibility for engine starting should have full knowledge of all procedures and regulations necessary to accomplish the procedure under unusual, as well as usual, conditions. (Unusual conditions might be starting engines with passenger stands positioned at an aircraft, with the jetway or jet bridge extended to the aircraft, or with the aircraft integral stairway extended.)

Figure 16-1. Those using hand signals to communicate with the flight crew should execute the signals in a clear, precise manner.

Photo courtesy of Northwest Airlines

All nose gear pins, main gear locks, pitot covers, control locks, wheel chocks, static ground wires, and power leads must be removed from the aircraft prior to departure.

As protection against fire, fire-extinguishing equipment should be in the area during every aircraft start. The size and type of equipment will vary with aircraft type and size. These units must be maintained in good working order, and all operators likely to be called on to operate them should be thoroughly trained in their use.

FOREIGN OBJECT DEBRIS

There are three basic types of FOD. Natural types of debris includes birds and ice. Human-made debris can be generated from anything on the ramp, including debris that falls off luggage and equipment. The third type can be generated from the operational activities of the aircraft.

Foreign object debris is a major safety concern for the ramp areas. It can be ingested by aircraft engines, possibly causing failure on takeoff or extensive damage to the engine that can lead to millions of dollars in damage. Unnecessary damage to aircraft tires also is a problem that can cost thousands of dollars to repair and replace.

Foreign object debris is generated by:

• Aircraft line maintenance activities on the ramp
• Baggage movement to and from the aircraft
• Cargo operations on the ramp
• Construction projects
• Careless littering on the air operation area by personnel
• Weather (heavy rains, freezing and thawing of pavement surfaces cause erosion or wind blowing debris across the ramp)
• Wildlife

Prevention techniques

The first step to preventing FOD is to form a committee consisting of the airport operator, airlines, and any other companies that service equipment or aircraft on the ramps. This will ensure a FOD program is properly structured and continually monitored at the appropriate level.

Airport operators and the airlines can take preventive measures to help prevent FOD. The most effective method is a ramp walk-down prior to aircraft arrival and after the plane departs. A ramp walk-down is lining up ramp personnel in a straight line, arms width apart, walking the entire gate area and picking up every piece of debris. Another method is using vacuum sweeper trucks to clean the ramp area on a daily schedule. For just picking up metallic debris, a bar magnet attached to a vehicle works well. To keep ramp personnel aware of the important role they play, consider signs posted at entrance areas regarding FOD or make a display case of FOD to be put in a prominent location such as the break room to remind people of the importance of the FOD program. FOD containers should be plentiful, conspicuously marked, and emptied on a regular basis.

AIRCRAFT LOADING AND UNLOADING

The well-being of passengers at the terminal ramp is of primary importance. The most common hazards for people in the ramp area are falling, tripping, and being struck by moving ramp equipment. In addition, bad weather causes hazards. The essential points in protecting passengers, as well as ramp personnel, are discussed in this section.

Passenger loading

The following specific points should help ensure passenger safety during boarding and disembarking:

• Portable passenger stairs or loading bridges should be properly positioned so there is no gap between them and an aircraft (Figure 16-2).
• After the portable passenger stairs or loading bridge has been positioned, the brakes and position locks should be set to prevent movement. The stair also should be properly set to prevent it from wobbling, bouncing or sagging.

- The maximum load capacity for passenger stands should not be exceeded. The load should be checked especially when the stand is used for group photos.
- Obstacles in the passengers' pathway from the plane to the gate also should be checked. (For example, check fueling hoses, ground wires, power cables, patches of ice or snow, oil or grease slicks, or water puddles. In addition, vehicles must be restricted from crossing the loading path during passenger boarding and disembarking.)
- The movement of other aircraft in the area should be checked before passengers board or disembark.
- Passengers should not be permitted on ramps while there is a danger of prop wash, engine ingestion, or jet blast.
- Movements of passengers always should be under control; passengers must not be allowed to roam free for both safety and security reasons.
- All employees should make sure that there is no smoking on the ramp-either by passengers or any airport or airline employee.

Figure 16-2. Passenger stands should be positioned so that there is no gap between them and the aircraft.

Photo courtesy of British Airways

Employee procedures at ramps

Because service employees are acutely conscious of meeting schedules, they may use undue haste during loading or unloading operations at a terminal ramp. Rapid, efficient work can be done safely, provided each member of the team does his or her job and uses proper methods and equipment. Any attempt to take shortcuts could result in delay, damage or injury.

The following employee guidelines should help to achieve a safe on-time operation.

- Each employee should understand and know how to carry out his or her particular function.
- Each employee should know the layout, function, and location of each section of the aircraft on which he or she works.
- It is important that only qualified operators be allowed to operate mobile ground equipment.
- A preplanned method of working on each specific aircraft will prevent confusion, mishandlings, and unnecessary rushing, all of which could lead to incidents and injuries.
- If an employee is in doubt as to how to perform a particular function, the employee should be encouraged to apply the maxim, "Don't guess. Ask someone who knows."
- Horseplay is a frequent cause of incidents. There is no room for such behavior either on an airport ramp or in the vicinity of an aircraft. All horseplay must be strictly prohibited.

Cargo handling

Employees who handle cargo probably are subject to more incidents and injuries than are other employees. This is because cargo handling usually requires physical effort on the part of the employee and the use of mechanized devices. Anyone involved in cargo handling should, therefore, be thoroughly trained in safe-operating techniques (Figures 16-3 and 16-4).

The following suggestions for handling materials can be used by employees as practical guidelines to handling cargo safely and efficiently:

- Do not pile or stack cargo too high. Avoid instability in cargo stacks.
- Pack all cargo firmly (either in the aircraft or on carts) to prevent its toppling.
- Use tarpaulins, cargo-restraining nets, side gates on carts, and other such items to prevent cargo from falling.
- Operate all mechanized units, such as cargoveyors and forklift trucks, in accordance with established instructions. Do not exceed the vehicle's rated load capacity. Carry loads centered and fully back against the fork carriage, and ask your supervisor if you have questions about the load. On an incline, always drive with the load up-grade. Always travel with the forks carried low, and do not carry loads that block your vision.
- Never attempt to lift, push or pull more than one's personal physical capabilities allow. When lifting, use the squat lift. Bend the knees, keep the back straight and keep the load close to the body. Do not twist the body and move feet towards the direction you want to move. If the load is large and/or heavy, get assistance to lift it. Use the stoop lift or assisted one-hand lift to lift small, light objects out of a container.
- Avoid wearing jewelry such as rings and identification bracelets. Such objects may catch on hooks, nails, buckles, locks, or straps and cause serious injury to fingers, hands or arms.
- All cargo should be slid rather than lifted into small spaces. Lifting may cause fingers or hands to become jammed between objects.
- Avoid handling cargo by the metal or plastic strapping that is frequently used to bind heavy or awkward shipments.
- Set all cargo down easily (rather than dropping it) to avoid injuries to the feet and toes, as well as to prevent damage to aircraft flooring and to shipments.
- Wear appropriate gloves, footwear and clothing.
- Exercise extreme care when entering the aircraft cargo areas from carts and tractors or when alighting from the aircraft onto vehicles. Entering and leaving aircraft is best done from stands.
- When handling live animal cargo, keep fingers and hands clear of the interior of cages or crates to avoid possible bites.

Figure 16-3 and 16-4. Efficient planning of the loading operation allows large amounts of cargo to be moved quickly and safely.

Photo courtesy of KLM Royal Dutch Airlines

Figure 16-4.

Photo courtesy of United Air Lines

- When a poorly packaged shipment is observed, proceed with caution.
- All contents should be prevented from spilling out and possibly causing injuries.
- Neckties should be avoided, but if they are part of a required uniform, make sure that ties are firmly clasped or tucked into their shirts to prevent possible catching in conveyor rollers or pinch points. A clip-on type of tie is desirable.

Weather

Unfavorable weather is as much to be reckoned with in conducting safe ramp operations as it is in conducting a safe flight. Rain, wind, and snow cause both employees and passengers to rush during conditions when footing and visibility are poor. Potential problems with ground equipment and aircraft also are compounded during inclement weather.

Foul-weather clothing serves to keep ramp operators comfortable and reduces the inclination toward undue haste. In addition, employees working during poor weather conditions must recognize the need to use extreme care both in their physical movements and in the operation of portable and mobile service equipment on ramps.

Aircraft and support equipment require special care in bad weather. Some rules to follow are:

- Use protective devices, such as engine intake covers, cooler plugs and pitot covers as weather conditions warrant.
- Keep footholds such as passenger boarding stairs, running boards on tractors and trucks, and ladders and work platforms clear of ice and snow.
- Keep handrails on boarding stairs clean and dry to encourage passengers use during rain, snow or wind.
- Check the outermost portion of jet bridges frequently to ensure that the walking surface exposed to the weather between docking is clear of ice and snow or water.
- Keep vehicle lights and windshields clear of snow and grime to aid the driver's vision.
- Use foul-weather equipment on vehicles such as

windshield wipers, defrosters, and frost shields. Such equipment should be maintained properly.

All possible steps should be taken to protect passengers from the elements. Covered passenger walkways are ideal. However, when these are not available, umbrellas or closed transportation from holding rooms to aircraft are helpful. During unfavorable weather, boarding and disembarking passengers should be closely supervised to prevent incidents.

Good maintenance of ramp surfaces is essential to employee and passenger safety, as well as to safe vehicle operation. Drainage systems should be maintained to carry away runoff from rainfall. Removal of snow and ice from ramps must be given high priority. Material used to thaw ice and snow on walkways and ramps must be noncorrosive so as not to damage aircraft and equipment.

GROUND EQUIPMENT MOVEMENT

Ground support mobile equipment operates as a vital adjunct to flying activity. Efficient control of ground traffic is necessary both to avoid equipment damage and personal injury and to run a safe, on-time operation.

Organizing vehicular traffic

Airport ramps were designed for aircraft, with vehicular traffic as a secondary thought. Without a coordinated vehicle control plan, there is great potential for incursion of vehicle traffic onto aircraft taxiways and runways. To avoid confusion, vehicle traffic flow, signage, and marking on the ramp should reflect local community traffic rules, regulations, and enforcement.

Regulations for airport vehicular traffic should cover the equipment operated on company premises and on airport roadways. For effective control, all groups concerned (airport management and airfield tenants) should coordinate efforts, agree on necessary rules and regulations, and accept responsibility for adherence and enforcement. These ends can best be achieved by establishing a local airport traffic committee consisting of airport management and tenant representatives.

These should be the prime activities of the airport traffic committee:

- Establish well-marked and controlled-traffic roadways, where possible.
- Post speed limits. Place signal lights and signs to indicate "no parking," "slow" and "stop." Designate authorized parking areas and traffic lanes.
- Develop methods for enforcement of the established traffic rules.
- Periodically review the types and numbers of vehicles that use airport roadways and aircraft apron areas to determine whether there are needless traffic lanes.
- Review traffic rules and procedures periodically and modify them as conditions change.
- Review incident reports and modify rules, where possible, to eliminate future incidents of a similar nature.

After the operating requirements and driving rules recommended by the airport traffic committee have been completed and approved by airport administration and management of the various tenants, the committee should publish and distribute the documents to all tenants and vehicle operators.

Operating requirements

Everyone who operates a motor vehicle on any airport service roadway or ramp area should carry a valid driver's license at all times when on duty. Some airports and/or tenants may issue permits to their own employees. If so, the airport administrator should have the authority to revoke such permits.

An airport driver's license should list the equipment that the holder is qualified to drive. To assist police officers, safety patrol officers, and others in authority, authorized drivers should be identified readily by a badge, uniform, armband or other conspicuous means established by the airport traffic committee.

Tenants and their employees should be required to maintain vehicles in good working order. Especially important are the horn, headlights, rear lights, rearview mirrors, brakes, reflectors, steering mechanism, turn indicators and windshield wipers.

All motor vehicles should display lights from one-half hour before sunset to one-half hour after sunrise, and at all other times when the airport obstruction lights are on or when there is not sufficient light to make persons and vehicles discernible at a distance of 150 m (500 ft).

Every vehicle authorized to operate on any public cargo ramp and apron area, public loading area, or public passenger ramp and apron area should display identifying symbols or numbers that can be clearly identified. Enforcement of this rule should limit field traffic to regular tenant vehicles and to other authorized vehicles such as supplier trucks and government agency vehicles.

Any motor vehicle authorized to operate in an aircraft landing area during daylight hours should be made conspicuous and identified by a paint scheme, checkered flag or other such means. Vehicles operating in this area during hours of darkness should have an overhead, rotating red light. In addition, motor vehicles operating in the aircraft landing area should have a two-way radio if aircraft movements are controlled by radio (or they should be accompanied by a vehicle so equipped).

Driving rules

No vehicle should be operated carelessly or negligently, or at a speed exceeding the limit. The operator must not be under the influence of alcohol or narcotics.

Vehicle operators in public areas should comply with rules of the road as established by the airport or airline. Traffic lights, signs, or mechanical or electrical traffic-control signals must be obeyed, unless an authorized representative of the administrator directs otherwise (Figure 16-5).

Lane requirements and safe following distances should be observed by vehicle operators. The rule for following distances is at least a two-second interval between two-axle vehicles. Longer intervals are required (four seconds or more) for larger, multiple-axle vehicles. More distance yet should be allowed under adverse driving conditions and for vehicles with loads.

Operators of ground vehicles must understand that aircraft always have the right of way. Ground vehicles should pass well to the rear of taxiing aircraft. When maneuvering around aircraft parked in the public cargo ramp and apron areas, ground vehicles should not move faster than a normal walk. They should be operated even more slowly on wet pavement, in congested quarters, and under other adverse conditions.

When an aircraft is parked on a public passenger ramp and apron area, the space between the aircraft and railings, fences or passenger loading gates should not be used as a vehicle thoroughfare.

Vehicle operators must know and obey control tower signal lights. Tower signal lights and their meanings are:

- **Steady red light:** Stop.
- **Flashing red light:** The vehicle is in a dangerous location and should be removed.
- **Steady green light:** Clear to proceed.
- **Flashing white light:** Return to origination point.
- **Alternating red and green lights:** Exercise extreme caution.

When the airport includes a localizer building used in conjunction with the instrument landing system, vehicle operators should be familiar with its location. Flashing red lights are installed at both ends of such buildings. When the control tower turns these lights on, all vehicles should stop behind the yellow lines on the ramp on both sides of the instrument runway. Vehicles should not proceed until cleared by the tower or other authorized traffic control personnel.

Vehicles should enter restricted operational areas only with the specific permission of the administration. In

addition, before entering or leaving restricted areas such as runways, taxiways, and fuel storage areas, the operator of a vehicle may be required to get permission from the control tower or other authority each time. Moving aircraft always have the right-of-way.

In general, tank trucks should be forward of the aircraft wing when overwing fueling is in process and far enough away from the aircraft to avoid damage in case of emergency. Trucks should be positioned to permit rapid removal in the event of an emergency and should not be blocked by other equipment. (Fueling is discussed in detail later in this chapter.)

All emergency vehicles should be given the right-ofway during an emergency. All other vehicles not needed for the emergency may proceed only after the operators have determined that they are not entering the emergency area or the path of emergency equipment.

Operators should not park a vehicle, or leave it unattended, in these public areas: landing ramp and apron, passenger ramp and apron, cargo ramp and apron, or aircraft parking and storage. Exceptions can be designated only by administration.

Figure 16-5. Airport ramps can become congested with ground service vehicles. Operators should strictly observe driving rules established by the airport.

Photo courtesy of JAL

Vehicles for hire should not load or unload passengers at any airport locations other than those designated.

AIRCRAFT GROUND HANDLING

The following information will guide readers in achieving a high degree of safety to personnel during aircraft pushback, taxiing and arrival/departure procedures.

Headset operator (aircraft pushback)

Due to the increasing number of fatalities and serious injuries that have occurred during this operation (caused by the operator being struck by the aircraft nose wheels and/or pushback tractor), it is strongly recommended that this operation be conducted from the comparative safety of the pushback tractor. This can be achieved by the use of the tractor interphone system or an alternate method using an external hardwire cord run from the aircraft to the pushback tractor.

Position of headset operator. The headset operation can be conducted by the tractor driver, who has the real control of aircraft movement and requires the fastest voice access in an emergency situation, such as an impending jackknife, or tow bar failure.

Alternatively, the headset operator may be positioned as a second person in the tractor, seated on an approved seat facing the direction of the pushback travel. In this method the tractor driver also should be on a live headset, and able to hear the headset operator and the flightdeck interphone voice communications.

The relocation of the communication person to the pushback vehicle, which is already a standard practice in some airlines, would virtually eliminate the necessity for anyone to walk in the danger zone during pushback.

Cordless headset systems. A number of manufacturers are making positive progress in the production of effective cordless headset communications systems. This method, as the name suggests, would again enable the communication person to sit in the tractor or, if for any reason it was found necessary to walk instead of riding with the vehicle, they could do so keeping well clear of the danger zone.

Hand signals

On arrival and during departure of an aircraft, before and after the use of voice communication is established, the provision and use of hand signals is required. These signals must be clearly understood by the flightdeck and handling crews. It is recommended that these signals be developed in accordance with International Civil Aviation Organization standards. If headsets are not used, direct communication should take place between the flight deck crew and ground crew to confirm the use of hand signals.

During nighttime operation, illuminated wands should be used. The signal person must maintain visual contact with the flightdeck crews at all times, so that his or her identity is clearly recognized by them.

Visibility

To ensure maximum visibility it is recommended that aircraft ground-handling crews wear clothing/vests with good light-reflective capabilities. Ideally, both the upper and lower parts of the body need to be clearly visible. This is particularly important during the hours of darkness, during poor weather conditions, or while working in areas where visibility is restricted.

It is essential that, before pushback operations commence and during the entire aircraft departure sequence, only authorized staff should be in the immediate vicinity Their location should be made known to the headset operator. Wing walkers must be in such a position so as to establish visual communications with the headset operator should the need so arise.

Procedures

Written procedures should be produced and made available to all members of aircraft handling crews. These procedures should give detailed guidance on aircraft arrival and departure procedures.

NOTE: ALL PROCEDURES SHOULD BE COMMUNICATED BETWEEN THE GROUND HANDLING COMPANY AND THE AIRLINES THEY SERVE.

Training

All handling crews must receive thorough training before participating in any aircraft arrival or departure.

The headset operator must receive particular training in pushback operations, with emphasis on voice and hand signal communications with the flightdeck, tractor driver, and wing walkers.

Refresher training must be carried out on all aspects of ground handling at predetermined intervals.

NOTE: THIS GUIDELINE REFERS TO ROUTINE OPERATIONS, SUCH AS IRREGULAR OPERATIONS. IRREGULAR OPERATIONS (IROPS), SUCH AS SEVERE WEATHER, REQUIRE ADDITIONAL PLANNING AND TRAINING.

PARKING

Parking areas for ground support equipment should be located in ready proximity to the aircraft parking areas on the terminal ramp, as well as to the servicing positions around the parked aircraft. These points should be considered for safety in the arrangement of parking and service locations:

- Location of fuel overflow vents on aircraft in relation to the service vehicle position
- No parking of the ramp service vehicles under wing tip fuel vents
- Possibility of exhaust vapors from service units entering aircraft cabins
- Avoiding water spillage (such as from lavatory and water-service units) in employee or passenger walkways
- Precluding aircraft settling onto adjacent equipment as fuel, cargo, and passengers board
- Keeping exit ways clear for moving fueling vehicles in the event of fire
- Keeping a clear pathway from aircraft and passenger stands in case of emergencies
- Positioning stands, food carts, ladders, or cabin service vehicles and conveyors so they remain stable in high winds
- Clear access for cargo tractor trains and other vehicles
- Clear access to fire extinguishing equipment
- Working room for aircraft service personnel
- Free and clear access to the aircraft for emergency vehicles

SERVICING AIRCRAFT

Aircraft servicing operations include fueling, providing passenger water and lavatory service, and deicing. Fueling is one of the most potentially hazardous servicing operations in the airline industry.

Millions of gallons of aircraft fuel are handled each year in servicing aircraft. Most of this fuel is handled safely because the potential fire hazard is recognized and the necessary precautions are taken. An understanding of fuel characteristics is important as a basis for establishing procedures to prevent fires, injuries to employees and the public, damage to aircraft and ground support equipment, and possible in-flight engine trouble.

This publication and others dealing with fuels (such as those of the National Fire Protection Association, petroleum companies, government agencies, and insurance companies and associations) define concepts necessary for a knowledgeable approach to safe fuel handling. Some basic information follows.

Three essential requirements must be satisfied before a liquid fuel can be burned.

1. The fuel must be converted to a vapor.
2. The fuel vapor must be mixed in the correct proportion with air (fuel-air ratio).
3. A source of ignition must be present.

Properties of fuels

Volatility: The tendency of a liquid to vaporize or change into a vapor.

Flash point: The minimum temperature at which

a fuel gives off vapor in sufficient concentration to form an ignitable mixture with air near the surface of the liquid. Ordinarily, the higher the vapor pressure (the pressure exerted by a volatile liquid, under any of the equilibrium conditions that may exist between the vapors and the liquid) of a fuel, the lower its flash point.

Flammable limits: Gases and vapors of flammable liquids have a minimum concentration of vapor or gas in air below which propagation of flame does not occur on contact with a source of ignition. This is known as the lower flammable limit (LFL). There is also a maximum proportion of vapor or gas in air above which propagation of flame does not occur. This is known as the upper flammable limit (UFL).

For example, a gasoline vapor-air mixture with less than approximately 1.0 percent of gasoline vapor is too lean, and propagation of flame will not occur on contact with a source of ignition. Similarly, if there is more than approximately 8 percent of gasoline vapor, the mixture will be too rich to ignite. Other gases, such as hydrogen, acetylene and ethylene, have a much wider range of flammable limits.

Flammable limits are determined at pressures of one atmosphere. Thus the range will increase with an increase of temperature or pressure, the UFL being more influenced than the LFL.

Flammable range: The difference between the lower and upper flammable limits, expressed in terms of percentage of vapor or gas in air by volume. It is sometimes referred to as the "explosive range."

For example, the limits of the flammable range of gasoline vapors are generally taken as 1.4 to 7.6 percent – a relatively narrow range. A mixture of 1.4 percent gasoline vapor (and 98.6 percent air) is flammable, as are all the intermediate mixtures up to and including 7.6 percent gasoline vapor (and 92.4 percent air). The range is the difference between the limits, or 6.2 percent.

Liquid turbulence: Agitation during tank filling that increases the fuel evaporation rate and vapor release. Because the flammability range is reached at a lower temperature than if no turbulence were present, the ignition hazard is increased.

Fuel misting: Fuel misting, or atomizing, changes the ignition hazard of the fuel. Any fuel is more readily ignitable if it is atomized or broken up into extremely minute particles. It is this characteristic, for example, that provides the ease with which home oil burners can be ignited. Although the temperature may be low, the fact that liquid takes on the form of a very fine spray increases its ignition capability. Therefore, a pinhole developing in a hose during fueling operations could create a fuel mist, which can be an extreme fire hazard.

Ignition sources

Static electricity: Because static spark discharge is a constant threat to safe fueling, proper bonding and grounding are essential. When two dissimilar materials make physical contact and are then separated, a charge of static electricity is almost always produced. Static electricity is generated when pumping any fuel through a system, such as a hose or pipe. The amount of static electricity generated increases linearly with the rate of flow. High fuel flow rates are normal today, with several thousand gallons of fuel often transferred in a relatively short period of time.

Static electricity can be generated by pumping fuel through a service hose, by allowing it to fall through the air from a filler spout into a tank, or by draining it from a tank or line into a container.

A charge may accumulate on an aircraft during flight or on the ground. Rain, snow, ice crystals, or dust blowing across the aircraft can create a heavy charge of static electricity. A passing cloud also may stimulate a charge. A servicing vehicle, like any other rubber-tired vehicle, may become charged too. Static also can be collected by induction from an electrically charged atmosphere.

Like water, static flows to points of lower potential. If the metallic structures on an airplane are bonded (or connected) electrically, the flow will continue until the potentials are equal.

Static electricity flows along the easiest path, just as lightning follows a highly conductive lightning rod and cable into the ground. If no easy path is provided, a charge builds up. When a charge is great enough, an electric spark jumps the shortest or most vulnerable gap. Often, this spark has sufficient energy to ignite flammable vapors.

CAUTION: STRAY ELECTRIC CURRENTS ORIGINATING FROM IMPROPERLY MADE CONNECTIONS BETWEEN THE GROUND POWER UNIT AND THE AIRCRAFT, OR FROM ELECTRICAL SHORTS IN ELECTRIC POWER SOURCES, ALSO MAY PROVIDE A SOURCE OF IGNITION.

Smoking: Matches, cigarette lighters, and other smoking materials should not be carried by any employees engaged in fueling operations. The no-smoking rule should be enforced rigidly. Because fuel vapors may settle and travel long distances along the ground, they can create a hazard even a distance away from the fuel source. If passengers remain aboard during fueling operations, there should be a qualified attendant in the cabin to enforce the no-smoking rule and to direct evacuation if it should become necessary.

Excessive fuel odors, or any of the other hazardous conditions noted, should cause fueling to cease until the condition is corrected. Passenger loading steps or loading bridges should remain in position during fueling, and an unobstructed exitway should be maintained for those on board to use in the event of emergencies.

Open flames: Open flames are employed sometimes in field maintenance work and as signal lights. Fueling should not be done while the following are within 15 m (50 ft) of an aircraft:

• Flareports and similar open-flame lights
• Welding or cutting torches or blowtorches
• Exposed flame heaters (including portable gasoline or kerosene heaters).

There should be no fueling while the aircraft engines or aircraft combustion heaters are running. This includes wing and tail deicing heaters.

Sparks: Electrical circuits may arc when connections are made; arcing also can occur from faulty equipment. The following precautions are recommended to prevent electric sparks:

• Aircraft batteries should not be installed or removed during fueling operations.
• Battery chargers should not be connected, disconnected, or operated during fuel servicing.
• Aircraft ground-power units should be located as far away from fueling points as practical; they should neither be connected nor disconnected during fueling.
• Care should be taken when using electrically powered tools or other such equipment in or near aircraft during fueling. Battery-powered tools appropriate for use in this environment should be considered.
• Aircraft electric switches that control units in the wing or tank areas and are unnecessary to the fueling operation should not be operated during fueling except in emergencies.
• Aircraft radios and radars should be off.
• Fueling should not be conducted within 30 m (100 ft) of energized airborne radar or within 90 m (300 ft) of energized radar installations.
• Photo flashbulbs should not be used in the immediate vicinity.
• Flashlights used near fueling points should be listed by Underwriters Laboratories for use in such locations.

Ramp vehicles: Improperly maintained vehicles present a fire hazard during fueling operations. Vehicle problems such as electrical system shorts, arcs across terminals, sparks from exhausts and backfires can ignite flammable vapors and should be corrected when first noticed in the vehicles.

No vehicles other than those performing aircraft servicing functions should be allowed within 15 m (50 ft) of the aircraft during fueling.

Footwear: Employees working near the aircraft should be discouraged from wearing shoes or boots with metal nails, hobnails, metal cleats, or plates on the sole or heel. Such footwear creates sparks when it makes contact with other metal or concrete, and damages wing surfaces and floor coverings.

Neighboring turbine aircraft: Fueling should not be carried on within 45 m (150 ft) directly downstream from the tailpipe of an operating turbojet engine, nor within 22.5 m (75 ft) directly downstream from an operating turboprop.

Fueling operations

Positioning of fuel servicing vehicles: A clear path should be maintained to permit rapid removal of fueling vehicles from the aircraft in an emergency.

Flow rates: When the aircraft is being fueled, the maximum rate of gallons per minute as outlined in the aircraft operations manual should not be exceeded. Nozzle pressure should not exceed 345 kPa (50 psi).

Overwing fueling from tank vehicles: This operation requires precautions against static electricity:

- A grounding cable should be connected from the service vehicle to a satisfactory ground.
- A grounding cable should be connected from the ground to an approved point on the aircraft.
- A bonding cable should be connected from the service vehicle to the aircraft.
- A bonding cable should be connected from the fuel nozzle to the aircraft before the tank cover is opened. This bond must be maintained until the fueling operation is completed and the fuel tank cover has been closed.
- Disconnections should be made in the reverse of the order stated above.

When mobile dispensing carts are provided with hydrant fueling systems, they should be grounded and bonded just as mobile fuel trucks.

Overwing fueling from hydrants, pits or cabinets: This operation requires the following grounding:

- A bonding cable should be connected from the grounded dispenser to the aircraft. A conductive fuel hose is not satisfactory bonding.
- A bonding cable should be connected from the fuel nozzle to the aircraft before the tank cover is opened.
- When disconnecting, reverse the order.

Overwing fueling from drums or cans: This procedure requires proper bonding and grounding connections, as do other methods. Drums should be placed near grounding posts to facilitate this. Connections should be in the following sequence. Before tank covers are opened, connect:

- Drum to ground
- Ground to aircraft
- Drum to aircraft
- Nozzle to aircraft

When disconnecting, reverse the order.

When fueling is done from cans directly into the aircraft (an uncommon procedure that should be used only in an emergency), a funnel and a chamois may be the only way to filter the fuel before it enters the tanks. Because this procedure is likely to increase the charge of static electricity and the possibility of sparks, use the following precautions.

Connect:

- Ground to aircraft
- Can to funnel
- Funnel to aircraft before tank cover is opened

Under no circumstances should plastic buckets or nonconductive containers be used for fueling.

Underwing fueling: This procedure reduces the chance of ignition of flammable vapors by an electrostatic spark. The integral wing fitting and mating hose connections are completely closed and, therefore, automatically bonded. The fuel does not flow until the attachment is completed, and a "splash filling" is avoided. However, grounding and bonding of the aircraft and equipment as previously outlined (except for the nozzle-to-aircraft bond) are still necessary to protect against static electricity and stray currents (Figures 16-6a and b).

Unusual terrain: When refueling is on ice, sand, desert terrain, or wherever else it is difficult to secure a satisfactory grounding, a compromise procedure is necessary. Because draining the static charge becomes impossible without an adequate ground, the potentials must be equalized to prevent a dangerous sparking discharge. Strict bonding practices must, therefore, be followed.

Lightning storms: Extreme caution should be observed during electrical storms, including suspending fueling operations during severe disturbances.

Indoor fueling: Fueling and defueling can be done indoors if guidelines provided by the NFPA are followed. (See References.)

Fuel spills: Prevention of spills entails considerable care on the part of every employee involved in the fueling operation. Fueling equipment must be maintained meticulously. There must be tight observance of each detail of the fueling procedures.

In the event of a fuel spill, emergency shutdown procedures should be activated at once. All fuel spills, regardless of size, must be regarded as a potential source of fire. Specific action to be taken will depend on the size and location of the spill.

Passenger water and lavatory services

Water servicing regulations and water service equipment handling procedures should conform with sanitary requirements of the appropriate health authority. Careful attention should be given to water spillage, which can be hazardous. Water freezing can create hazardous conditions on ramps and within the serv-

icing areas of an aircraft. Precautions must be taken to make sure aircraft are heated properly and that water freezing is prevented. Electric water heaters require attention; they can burn out or suffer other damage.

In many instances, lifting devices, platforms, or access stands are used by operators to reach the water service areas of aircraft. Operators on such devices should use extreme care to prevent falls. They should wear safety harnesses and lanyards where conditions warrant them.

Operators of water service vehicles and other mobile water service equipment should be careful to prevent injury to themselves or damage to either the equipment or the aircraft. For example, rough handling of latching devices on water service access panels should be avoided to prevent damage to the aircraft.

For the most part, lavatory servicing requires the same precaution and care necessary for water servicing and hose disconnection. In addition, attention should be given to the relationship between the locations of lavatory service panels and cabin air intakes. Sometimes, odors resulting from servicing can be drawn into the aircraft cabin through these intakes. If this occurs, intake fans should be shut off, as a temporary expedient. Engineering modification should be sought as a permanent solution.

Local health regulations pertaining to waste disposal must be followed closely to prevent contamination or spread of disease. For the same reason, lavatory service equipment should be kept in sanitary condition and good repair. During lavatory servicing, minimum protective equipment should include a faceshield and impervious gloves.

Deicing operations

The presence of water, ice, and snow on control and wing surfaces can have a very serious effect on aircraft operation. The aircraft's rapid rate of climb to – and descent from – high altitudes increases the possibility of icing conditions in any season.

When freezing or near-freezing temperatures are likely

Figure 16-6a. Underwing fueling from a hydrant.

Photo courtesy of Qantas Airlines

Figure 16-6b. Underwing fueling from a tank truck.

Photo courtesy of United Air Lines

to be encountered, an aircraft should not be dispatched with water or wet snow on its flying surfaces, control surfaces, hinges, or any other part likely to be adversely affected by such an accumulation. When it is known that a hazardous condition could result from accumulation of moisture, it is extremely important to keep the aircraft moisture-free.

Ice and snow that have accumulated on aircraft and that cannot be swept off by a broom, a rope, or muslin strips should be removed by an approved deicing fluid. Deicing fluid can be applied with mops, cloths, hand pumps, or power sprayers. Those applying the fluid should wear appropriate personal protective equipment.

It is important that the fluid be applied as uniformly as possible. No fluid should be permitted to enter any intakes, vents, or openings in the aircraft engines.

On aircraft that must be refueled over the wing, deicing should not be started until the fueling is completed (Figure 16-7). Deicing fluid on upperwing surfaces presents hazardous footing for employees engaged in fueling.

Deicing fluid should not come into contact with the exhaust stacks or other hot engine components. Although the fluid has a relatively high flash point, it still can be ignited.

Deicing fluid should not be used to remove ice or snow from props or prop-spinners. In many cases, deicing is accomplished electrically from the flight-deck. (A cloth dampened with deicer fluid may be used; however, manufacturer's procedures should be checked.)

Special attention should be given to aileron and elevator sealing strips, hinges, airseal diaphragms and flap telescoping rods. These areas may become congested with snow or slush following ice removal.

In some instances, a long aluminum tube with one end covered with a rubber hose can be rubbed across various surfaces to remove ice. Under no circumstances should the aircraft be hit or pounded with these tubes in an attempt to break up ice concentrations.

All mechanized deicing equipment should be operated with extreme caution because of the close proximity to the aircraft. Aerial baskets – baskets attached to a boom used to raise operators to a work area – should be treated with extreme caution and operated precisely according to authorized methods. These units can be hazardous if there is not complete coordination between the person in the basket and the one at the driving controls.

Aircraft also may be deiced in heated hangars. When this is performed, it should be ascertained that the aircraft is dry before moving it outside; otherwise, any moisture remaining after the melting process might refreeze. However, the aircraft may be moved outside without complete drying if a coat of deicing fluid is applied prior to removal from the hangar.

NOTE: DEICING FLUID IS CONSIDERED A HAZARDOUS MATERIAL, AND SHOULD BE USED, COLLECTED, AND RECOVERED IN ACCORDANCE WITH FEDERAL AND LOCAL PROCEDURES. WORKPLACES, OPERATORS, NON-OPERATORS, AND RESOURCES NEED TO BE APPROPRIATELY PROTECTED.

COMMUNICATIONS IN RAMP AREAS

One of the most important safety procedures in parking or moving aircraft on ramps involves proper signaling procedures. An aircraft, because of its weight and size, is a difficult object to stop, start, and maneuver in a restricted area.

Standard hand signals, with colored paddles during daylight and light wands in darkness, should be used by every operator (Figure 16-8).

Supervisory personnel must ensure that only qualified and competent employees are allowed to signal an aircraft. Signals should be given to the captain of the aircraft by only one person at any time.

Most operators require passengers remain seated until the aircraft has come to a full stop – a difficult rule to control. Therefore, an aircraft should be signaled into the parking area in as smooth a manner as possible to preclude sudden stops and starts or extremely sharp turns. The signaler is essentially the eyes and ears of the flight crew while the aircraft is taxiing into or out of a parking area. The captain of an aircraft relies on the signaler to guide the plane safely, smoothly, and efficiently.

While an airplane is standing after it has been parked but before it is to begin taxiing, interphone or two-way radio communication has advantages over hand signals or the voice.

Interphones or headset systems. Various types of ground-to-aircraft interphones or headset systems are available. Everyone assigned to use such equipment should fully understand its function and their responsibility. Interphones or headset systems should be well maintained to ensure correct operation at all times. Intermittent communications with breaks can be misunderstood easily and, as a result, potentially dangerous conditions can be created. Voice communications procedures should be strictly followed. Unorthodox commands or acknowledgments must not be tolerated.

Two-way radios (mobile or walkie-talkie). Mobile radio or walkie-talkie systems provide an efficient method of ground-to-flightdeck communications, provided that personnel are fully aware of the operating procedures and that the equipment is properly maintained. Such systems should be monitored at all times to maintain complete communication. Correct voice communications procedures should be closely followed, and horse-play or unauthorized conversation should not be tolerated.

PROTECTION AGAINST JET BLAST AND NOISE

Jet blast on terminal ramps is a hazard to employees, passengers and equipment. Adequate ramp surfaces and ramp procedures can help reduce the possibility of jet blast damage.

Reducing jet blasts
To reduce jet blasts, concrete ramps in good condition probably make the most satisfactory surface. Airline and airport management should work together to ensure proper maintenance of ramp surfaces. Concrete ramps that have begun to deteriorate by sinking, cracking and crumbling generally require that an aircraft produce additional thrust to compensate for their bad condition.

A macadam surface may be less satisfactory than a concrete one because it has a tendency to "dish out" under the main aircraft gear in the hot sun; as a result, the aircraft needs to produce additional thrust for breakaway.

Poor weather conditions, regardless of the type of ramp, can cause increased jet blast. Ice and snow on the ramp surface will require greater thrust from the aircraft engine to begin initial pullout. Airport authorities and tenant operators, therefore, should maintain ramps as free of ice and snow as possible.

Ramp equipment should be stowed so that it is unaffected by blast. This is essential to avoid personal injury and property damage. Blast fencing can help reduce injury and potential damage.

Aircraft should be parked to prevent any blasts from being directed at galleries or balconies where large gatherings of the public can be expected. Monitoring pilot techniques and surrounding conditions can help reduce blasts and their possible damaging results.

Figure 16-7a, b and c. Operators in aerial baskets deice an aircraft.

Photo courtesy of Canadian Airlines

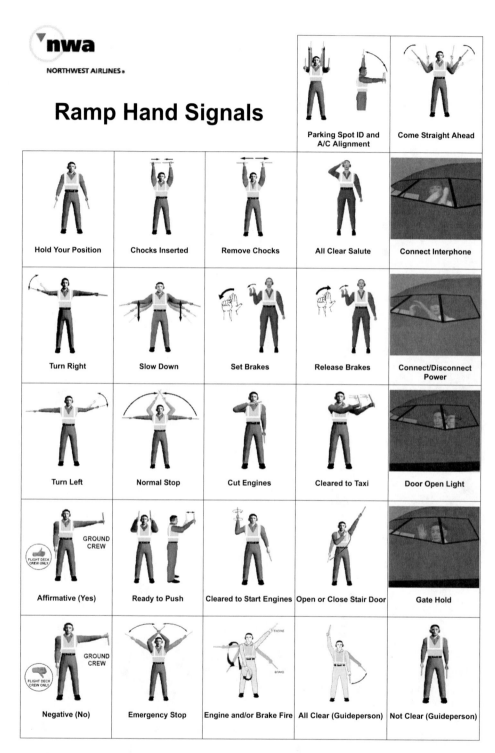

Figure 16-8. Basic hand signals used for communications in ramp areas.

Photo courtesy of Northwest Airlines

REFERENCES

American National Standards Institute. *Deicing/Anti-icing Fluid, Runways, and Taxiways, ANSI SAE AMS 1426A.*

National Fire Protection Association:

   Aircraft Fuel Servicing, NFPA 407, 1996.

Airport Terminal Buildings, Fueling Ramp Drainage, and Loading Walkways, NFPA 415, 1997

Aircraft Maintenance, NFPA 410, 1994.

Portable Fire Extinguishers, NFPA 10, 1994.

SEVERE WEATHER GUIDELINES

CHAPTER 17

Definitions
Tools for severe weather awareness
Severe weather forecasting
Local detection devices
Severe weather notification: Developing your plan
Means of notification
Being prepared
Recommended procedures: Using your plan
When severe weather and lightning is imminent
Appendix A: Airport weather warning (AWW)
Background
Procedures
Issuance criteria
Appendix B: Lightning safety
General information regarding lightning hazards
Principles of lightning protection
Lightning "safe" areas
Personnel safety Dos and Don'ts when severe weather is imminent

Severe weather, generally defined as high winds and lightning (see the "Definitions" section below), is a constant danger to all industries that conduct outdoor activities. The aviation industry is particularly affected because of:

- Open areas of airports
- Large, metallic pieces of equipment, including aircraft
- Numerous open-cab pieces of ground support equipment which are a part of the work area

This document outlines industry practices that can minimize the dangers associated with severe weather and the aviation workplace.

DEFINITIONS

- **High winds:** Sustained winds in excess of 40 mph (34.8 knots, 65Kmi/h) or gusts more than 50 mph (43.5 knots, 81km/h).
- **Lightning:** An electrical discharge between clouds and ground.
- **Primary alert:** Severe weather is within 10 miles (16 km).
- **Operational ban:** Weather is within 3 miles (5km)

and moving in the direction of your operation.
- **All clear:** Weather has moved beyond 3 miles (5km) and is moving away.

NOTE: HEAVY RAIN OR SNOW WILL NOT BE CONSIDERED IN THIS DOCUMENT BECAUSE THE EFFECT IS LESS SUDDEN AND LESS LIKELY TO CAUSE SUDDEN DAMAGE OR INJURY.

TOOLS FOR SEVERE WEATHER AWARENESS

Severe weather forecasting

Knowing when severe weather will hit your work site is a key element in prevention of injuries or damage. The following are possible sources of timely information regarding approaching severe weather:

- Internal weather forecasting
- Local TV and radio broadcasts
- Pilot reports
- Federal Aviation Administration (FAA) airport tower observations
- Ramp tower observations
- Local detection devices
- Weather service providers
- National Weather Service (United States):
- The National Weather Service provides bulletins about severe weather using a standard definition of severe weather. These bulletins should be monitored and the information from them communicated to your operations group.
- The Airport Weather Warning (AWW) (See Appendix A for more detail) is a specialized bulletin available from the National Weather Service in hub cities for the U.S.; other countries may have similar services. The AWW can be tailored to the specific needs of an aviation operation. Contact your local office of the National Oceanic and Atmospheric Administration (NOAA) or the U.S. Weather Service for more information.
- NOAA United States weather radio
- Internet sources (Use caution when getting weather information from Internet sites. The information may be unreliable or dated.)
- Intellicast.com
- Weather.com
- IWIN.nes.noaa.gov

- IWST
- County paging systems: some counties have paging systems for emergency management personnel. You can become a part of the notification group by contacting your local county emergency management authority.

Local detection devices

Automatic detection systems that track storms, count and locate each lightning strike, and determine the potential for lightning strikes based on atmospheric conditions are available. These systems may already be in place in some locations. Carriers or airport authorities may purchase them.

NOTE: SOME OF THESE SYSTEMS ARE PRONE TO FALSE POSITIVES; SOME PREDICT LIGHTNING POTENTIAL THAT MAY NEVER MATERIALIZE INTO ACTUAL STRIKES. THIS CAN CAUSE PERSONNEL TO LOSE CONFIDENCE IN THE WARNING SYSTEM.

These systems require human monitoring and human interpretation. Systems that combine several methods of detection along with visual observation are the most effective.

Handheld lightning detectors are small devices that can be used to give some indication that there is impending severe weather in the area.

A combination of some or all of these methods is probably the best one. Some experts consider quality local prediction systems the most effective tool.

SEVERE WEATHER NOTIFICATION: DEVELOPING YOUR PLAN

Means of notification

Notifying personnel that severe weather is approaching or imminent is a challenge. One system may not work in all locations or under all conditions. Regardless of the size of the airport, the airport authority, all airlines, vendors and the control tower must be actively engaged in whatever notification process is considered the most suitable. The following are some recommended means of notification:

- **Radio:** Good for small areas where workers perform their duties together and team leaders with radios can notify everyone.
- **Visual signals:** Lights on structures that indicate "Take Shelter Indoors" may be useful, but must be distinctive so they don't blend into the background or become confused with other lights. How these lights are activated and by whom must be clearly defined.
- Some airports use several different color lights: green for "all clear"; yellow for warning; red for take shelter.
- Some experts recommend a blue flashing light as best for single light systems because it is less likely to be confused with other lights.
- **Audible signals:** Horns or sirens may be used, but must be able to be heard above engine and equipment noise.

A combination of all the above is recommended.

Being prepared

Have a plan. Create a plan for daily operations and a plan for when severe weather is imminent. Work with airport authorities so each plan complements the other. Make certain vendor services and other airlines are included in and aware of the plan.

Several actions can be taken on a day-to-day basis to ensure minimal damage during a severe weather event. The following suggestions have proven effective. They can result in lesser amounts of damage when advance notification of approaching severe weather is provided or when a severe weather event occurs without warning.

- Regularly set and check brakes on all parked equipment.
- Clear debris from all areas.
- Keep only equipment currently being used in the area. The balance should be secured in holding areas away from the operation.
- Set up a hitching rail in equipment storage areas; chain up all unused equipment to the hitching rail when it is stored. This avoids the need to secure the entire equipment inventory every time a severe weather warning is received.

- Use a restraint system on container storage racks; habitually check the system to ensure it is working.
- Park bag carts against the building whenever possible; leave the curtains open when not in use.
- Chock all or at least more than one gear on aircraft.
- Pull loading bridges as far as possible away from overnight aircraft.

Know your aircraft:

- Some aircraft can be moored.
- Some aircraft have torsion links that should be connected.
- Some aircraft should have aft stairs down to stabilize the aircraft.
- Some aircraft must have aft stairs up.

RECOMMENDED PROCEDURES: USING YOUR PLAN

When severe weather is forecast:
Start of the day:

- Meet with ground operations, ground support, equipment, and maintenance managers to outline the forecast and review resources.
- Communicate to all personnel that the "Severe Weather Plan for (high winds, lightning or hail) is in effect."
- Notify dispatch, passenger service, and planning groups to expect delays.
- Assign a specific person to monitor the forecasts (if no one is already routinely assigned this duty).
- Communicate with airport authority for information about airport or runway closures.

Before severe weather arrives:

NOTE: HAVE A PLAN THAT INCLUDES A TIME STUDY TO DETERMINE HOW LONG IT WILL TAKE TO PERFORM ALL ACTIVITIES BELOW.

- Notify all personnel that an approaching severe weather event has been announced.
- Make decisions on which aircraft will continue boarding or deplaning passengers.

1. Secure aircraft

- Chock main gears and nose gear of all aircraft.
- Set aircraft brakes if possible.
- Secure all cargo nets.
- Close all cargo doors on aircraft not actively loading or unloading.
- Secure all aircraft cabin doors on aircraft not boarding or deplaning passengers.
- Close cockpit windows.

CAUTION: IF THE PASSENGER CABIN DOORS ARE SECURED WHEN THE APU/PACKS ARE OPERATING OR WHEN AN EXTERNAL CONDITIONED AIR SOURCE IS CONNECTED, THE AIRCRAFT CAN BECOME PRESSURIZED.

- Close all service panels.
- Lock control surfaces in accordance with aircraft maintenance manuals.
- Raise aft stairs on aircraft that do not use stairs for tail stands (up on DC-9s, down on B-727s).
- Secure torsion links on nose gears of aircraft that have them to prevent weathervaning with free moving nose wheels.
- Follow manufacturer's mooring procedures for aircraft that can be moored and where prior arrangements (mooring points in ramp) have been made.
- Follow manufacturer's procedures to add or transfer fuel on aircraft as a high-winds condition precaution on aircraft that carries this manufacturer's suggestion.
- Hook up tow bar and attach tugs if possible.
- If time permits and parking areas are available, move aircraft into hangars. If hangars are not available, consider remote parking aircraft to get them away from structures into which the wind can blow them.
- Face aircraft into the wind, if possible.

2. Secure loading bridges

- Retract ground power cords.
- Retract loading bridges as far as possible from aircraft.
- Lower loading bridges.
- Position loading bridges so they face the wind,

or close to the terminal, if possible. This can reduce the surface area in which the wind can blow them around.

- Set wheel in a direction to minimize bridge movement.
- Where available, position loading bridges in location for tie down and tie them down.
- Chock loading bridges and close all doors.
- Remove and empty all FOD containers.
- Remove stored ladders, de-icing poles and any loose equipment.

3. Secure ground support equipment (GSE)

- Secure or stow indoors all ramp equipment not in use. This includes chocks, cones, wands, FOD buckets, ladders, air conditioning hoses and any other loose items.
- Secure or stow indoors all luggage carts, containers, transporters and dollies; for those that must remain outside:
- Park them parallel and away from aircraft and set brakes.
- Disconnect strings (trains) so each conveyance is held by its own brake, or attach a vehicle to them to help hold them in place.
- Ensure all containers are locked on dollies or transporters with doors or curtains secured.
- Remove all empty, loose containers from area around aircraft; tie them together or to a firm structure, or store them indoors.
- Disconnect all ground support equipment from aircraft and store outside the path of possible aircraft movement if practical, and park parallel to aircraft.
- Secure work stands by chain to hitching rails, fences or other secure equipment.
- Put jackscrews down if so equipped.
- Lower high-lift vehicles and deploy stabilizers.
- Park all motorized equipment away from aircraft with parking brakes set.

CAUTION: DO NOT OPERATE WIDE-BODY LOADERS WHEN WINDS ARE EXPECTED TO EXCEED 40 MPH (34.8 KNOTS, 65KM/H).

4. Cargo

- Ensure notification of the severe weather alert has been conveyed to cargo personnel.
- Cargo personnel should:
- Discontinue delivery of shipments to aircraft.
- Discontinue delivery of shipments outside to be stored.
- Secure all cargo equipment as detailed in the Ground Support Equipment section above.
- Contact mail and freight facilities and ensure they are aware of approaching severe weather.
- Return unloaded freight and mail.
- Stow luggage in bag rooms if possible.

5. Aircraft cleaning

- Stow all supplies and equipment.
- Leave no equipment or garbage on loading bridge steps.
- Move vehicles away from aircraft parking areas.
- Stow hoses from lavatories (toilets) and water trucks.

6. Maintenance

- Stow all ladders, stands and other loose equipment in building or secure them.
- Park all vehicles as detailed in the "Secure ground support equipment" section, or remove them from the ramp altogether.

7. Vendors

- Notify all vendors about the severe weather event and ensure all vendors know your plan.
- Remove all loose or unsecured equipment.

8. Other airlines

- Provide a copy of your plan to other airlines.
- Notify other airlines of any unsecured equipment.
- Solicit other airlines' support in securing all equipment.

9. Facilities

- Ensure facilities personnel have been notified of the impending severe weather.
- Ask them to stand by for:
- Possible shutdown of power
- Possible need for facility repairs
- Close all doors that lead to the outside.
- Secure all dumpsters and trash bins.

10. Passenger safety

- Ensure all passenger service personnel have up-to-date information on severe weather alert.
- Be prepared to move passengers to safer areas in terminals, away from windows.
- Notify passengers at home of delays whenever possible. (Some passengers may delay coming to the airport, which may help ease congestion.)

If passengers have not started boarding an aircraft:

- Hold them in gate lounges.

If passengers are boarding:

- Stop the process and consult with the flight crew.
- Consider leaving on the aircraft those who already have boarded.

NOTE: THIS DECISION SHOULD BE BASED ON THE EXPECTED SEVERITY OF THE STORM AND THE SIZE OF THE AIRCRAFT.

- It may be wise to remove passengers from regional jets or smaller aircraft if there is time to do so safely.

If passengers are arriving:

- Move them into the terminal as quickly as possible.

11. Flight crews

- Ensure pilots have been notified that brakes should be set on all parked aircraft because of anticipated high winds.

12. Shift change

- Ensure all personnel coming on duty know that the severe weather plan is in effect.

When a severe weather forecast includes lightning:

- Ensure all personnel know of impending severe weather and that lightning is possible.
- Be prepared to stop fueling operations when storm is within 5 mi (8 km) and is moving toward the operation.
- Ensure all ramp personnel know to discontinue aircraft communication by headset when storm is within 3 mi (5 km).
- Be prepared to stop all ramp activity and clear the ramp when storm is within 3 miles (5 km).
- Ensure no one plans to use aircraft for cover.

WHEN SEVERE WEATHER AND LIGHTNING ARE IMMINENT

When storm is within 5 miles (8 km):

- Stop all fueling operations.

When storm is within 3 miles (5 km):

- Discontinue boarding and deplaning passenger.
- Close all aircraft doors.
- Stop all headset communications.
- Use the telephone only in an emergency.
- Evacuate the ramp.
- Advise arriving aircraft to hold short of gate or designated parking location.
- Ensure no one seeks shelter under any part of the aircraft, loading bridge, near light poles, fences or under trees.
- If a person is unable to get into buildings, seek

cover inside aircraft or inside an enclosed metal-bodied vehicle.

APPENDIX A: AIRPORT WEATHER WARNING (AWW)

Background

The Airport Weather Warning (AWW) is a U.S. National Weather Service (NWS) product designed to address weather phenomena that may adversely impact ground operations at airports. This information has proven useful to airport managers, fixed-based operators, airline ground personnel and others responsible for the safety of ground operations.

Ground decisions based on an AWW may include those associated with fueling delays during thunderstorms, de-icing frequency and other similar ground operations. AWWs are not intended for use by in-flight operations.

Procedures

An AWW is issued based on airport-specific weather criteria and/or the issuance of any NWS warning product that will affect the airport within a 5-mile (8 km) radius of the center of the airport complex – as agreed upon between a local airport management and the supporting WFO. A letter of agreement determines the weather elements, dissemination mechanisms and points of contact.

An AWW complements existing NWS warnings and forecasts. Airport officials are encouraged to refer to other NWS warning and forecast products, such as Terminal Forecasts (TAF), Short Term Forecasts (NOW), Zone Forecasts (ZFP), and other public watches and warnings. These also contain information about conditions that may impact ground operations. Use of an National Oceanic and Atmospheric Administration (NOAA) weather radio should be encouraged.

An AWW is written in a plain-language, free-text format. It typically includes the triggering phenomenon, the location and the start time (and end time as needed), and may include additional remarks.

Issuance criteria

Issuance criteria are established according to local airport requirements, and should be reviewed on an annual basis and updated as necessary. They include:

- Surface wind gusts of 40 knots or greater
- Cloud-to-ground lightning with 5 miles (8 km) and approaching the airport
- Thunderstorms with hail of half an inch (1.2 cm) or greater
- Onset of freezing rain
- Onset of heavy snow

APPENDIX B: LIGHTNING SAFETY

General information regarding lightning hazards

Lightning is an electrical discharge between differently charged regions within the cloud or between charged clouds to earth. A complete ground flash consists of one or more high-amplitude, short duration current impulses called strokes. Sometimes this is followed by long duration, low-amplitude currents.

The location of each lightning hit, as well as the magnitude of each flash, is a stochastic phenomenon. Lightning flashes generated within one isolated thunderstorm cloud can hit any point on the ground beneath the cloud. The distance between two successive flashes varies in the order of miles.

The principal effects of a direct lightning strike to a person or an object are electrical, thermal, and mechanical. A lightning strike to a remote object linked by conductive parts (e.g., power or communication cables) can cause electrical effects that are hazardous to sensitive electronic equipment or trigger an electric spark that is dangerous in high flammable areas.

As a rule of thumb, anyone who can see lightning and/or hear thunder is at risk. Louder or more frequent thunder indicates that lightning activity is approaching, increasing the risk for lightning injury or death. If the time delay between seeing lightning and hearing thunder is less than 30 seconds, seek a lightning-safe area.

NOTE: THIS METHOD OF RANGING HAS SEVERE LIMITATIONS DUE TO THE DIFFICULTY OF ASSOCIATING A BURST OF THUNDER TO THE CORRESPONDING FLASH OF LIGHTNING.

High winds, rainfall and cloud cover often are precursors to actual cloud-to-ground strikes, providing advance warning to take action. Many lightning casualties occur as storms approach because people ignore these precursors. Also, many lightning casualties occur after the perceived threat has passed. Generally, the lightning threat diminishes with time after the last sound of thunder but may persist for more than 30 minutes. When thunderstorms are in the area but not overhead, the lightning threat can exist even when it is sunny, not raining, or when clear sky is visible.

Remember that lightning may strike many miles from the edge of the thunderstorm cell. Acceptable downtime in ramp and other open-area operations must be balanced with the risk posed by lightning.

Principles of lightning protection

The purpose of lightning protection is to protect people, buildings or structures in general from lightning – to a certain acceptable level. There is no 100 percent protection level within technical and economical constraints.

Lightning protection is not aimed at preventing the formation of the lightning discharge. It is intended to prevent an object from being hit directly or affected by a remote lightning discharge. Therefore, lightning protection systems typically contain four components:

1. Air terminations: To provide interception-lightning protection

2. Down conductors: To convey lightning current to earth

3. Earth termination network: To deliver lightning current into earth

4. Over-voltage protection: To prevent hazardous potential differences while allowing normal operating potentials to exist

Lightning 'safe' areas

No place is absolutely safe from lightning threat; however, some places are safer than others:

- Inside a terminal building
- Inside a fully enclosed metallic vehicle
- Inside a safety shelter

Personnel safety Dos and Don'ts when severe weather is imminent
Do:

- Use towbar-less air tug for aircraft pushback.
- Use cordless headsets.
- Pay attention to lightning alerts.
- Take shelter when directed or when immediate weather hazards are seen.
- Stay a safe distance away from isolated, large, metallic structures.

Don't:

- Get off vehicles.
- Use headset connected to aircraft.
- Stay in open areas of under aircraft.
- Seek shelter under a tall tree.

WORKING GROUP MEMBERS

Chairman:

Jim Stephan – Delta Airlines

Members:

Michael Lueck - ABXAir Inc

Tom Dyce – Air Canada

Willy Pfister – SP Techinics, TEB

AIRCRAFT MAINTENANCE
OPERATIONS

CHAPTER 18

Aircraft maintenance in hangars and workshops brings together representatives of nearly every aircraft trade and skill, along with many other categories of support personnel. This concentration of personnel, the limits of space in which they work, the variety of activities they perform, and the size and weight of the aircraft and equipment involved create a significant safety challenge.

HANGAR OPERATIONS

For safe hangar operations continual attention should be directed towards: the main power systems of the hangar itself; good housekeeping and fire protection; safe work practices; and the use of protective clothing appropriate for the operation being performed.

Effective control of these factors can help reduce the risk of accidents.

General rules for hangar facilities

Housekeeping: Housekeeping can be defined as "a place for everything and everything in its place." In hangars, housekeeping is a continuous operation. The very nature of the work produces oil spills, used parts, dust, grime, and debris on work surfaces. A sound housekeeping program should result in cost savings by reducing accidents and reducing the potential for fire. Here are some basic housekeeping rules for hangars:

- Keep floors clean and free from oil, grease, fuel and trash at all times.
- Keep aisles, walkways and exits clear.
- Establish a ready access clear zone in accordance with applicable standards on all sides of a hangar, shelter or nose dock. No storage, vehicle or aircraft parking, or other building of any type should be allowed in this area.
- Avoid concentrations of equipment, toolboxes and other materials around aircraft undergoing overhaul, because these items can form obstacles.
- Mark areas of operation and protective devices.
- Guidelines for positioning aircraft should be clearly visible and kept intact.
- Dispose of oily rags, empty oil cans, and other combustible waste in metal containers with self-closing lids. Place the containers in convenient locations in the work area, and empty them on a regular schedule.

Electrical equipment. Both fixed and portable electrical equipment in hangars and around aircraft should be of an approved type according to your local, state or federal authority; or your country's national standard body, such as the National Electrical Code (NFPA 70; see References).

All electrical equipment in hangars should be inspected continuously to detect defects that may have arisen as a result of use or misuse. Electrical control panels and switchboxes must be accessible at all times.

Compressed air systems. Compressed air lines in hangars should have approved fittings, without makeshift modifications. Hose lines should be approved for the pressures involved. Hoses should be securely connected in place to eliminate both tripping hazards to people and damage to hoses. The operator should check hoses before use to make sure they are securely connected to the proper pipe outlets. Air hoses should receive periodic inspections. Defective hoses should be replaced immediately. High- and low-pressure chucks for servicing systems should not be interchangeable. It should be impossible to interchange low-pressure systems with a high-pressure supply.

Aircraft painting and cleaning. Painting and cleaning operations usually involve the use of flammable and/or toxic corrosive substances. Every effort should be made to protect employees from exposure to hazardous materials. Employees shall be provided with proper protective equipment and should use it. Other necessary controls must be used as suggested by the Material Safety Data Sheet (MSDS) available from the manufacturer, distributor, or supplier of the substance. In the United States, also check the requirements of the OSHA Hazard Communication Standard mandated by the Occupational Safety and Health Act of 1970. Other regulations and standards may exist outside the United States.

Whenever possible, the least flammable and least toxic materials should be used. Flammable solvents should not be used in the cabins of aircraft.

All hazardous materials should be properly identified by:

1. Name
2. Health hazards they pose
3. Safety hazards they pose
4. Control measures for them

Employees exposed to these materials should be provided with training for each potentially hazardous material.

Hangar door operations

Hazards associated with the operation of aircraft hangar doors have caused several serious injuries, including fatalities. Operators must be fully aware of these hazards and given instructions on the safe operation of hangar doors, including proper maintenance procedures. No employee should be permitted to operate a hangar door without receiving training.

Aircraft docking

Whenever possible, the aircraft should be defueled prior to entry into a hangar.

The steps in positioning or docking an aircraft are:

* Make sure the hangar doors are fully open.
* Warn operators of an approaching aircraft by posted signs or other means.
* Make sure qualified operators are at the aircraft controls during the entire operation.
* Post observers at each wing tip and at the tail to warn the operator of the tug of unseen hazards.
* Make sure all aircraft enter on the proper guidelines on the hangar floor.
* Position aircraft and attach ground plugs before closing the hangar doors. (Wheels also should be chocked as required.)
* Draw dock structures into position before removing warning signs.
* Make sure all movable structures are locked into position, guardrails in place and utilities connected.

The steps in removing, an aircraft from the docks are:

* Remove all equipment and personnel from a dock.
* Disconnect grounds, power, air pressure and other utilities.
* Remove movable structures around the aircraft.
* Open the hangar doors fully.
* Position a qualified operator in the aircraft and at the controls.
* Connect tugs, picket the aircraft at both wing tips, and place one person outside hangar doors to guide the tug operator. This will ensure no other

approaching traffic or personnel are in the path of the tug and the aircraft.

- Remove chocks.
- Check that the aircraft is guided clear of all obstructions and position on ramp.
- Place chocks in position.
- Set aircraft brakes before disconnecting towbars.
- Close the hangar doors.

Aircraft fuel systems

Aircraft fuel system service and maintenance personnel should strictly observe the provisions of applicable maintenance manuals. In maintaining aircraft fuel systems, consider the two basic hazards: fire and health.

Fire hazards. Aviation gasoline (avgas), jet A, and JP fuels are classified as flammable or combustible liquids, depending upon grade. All precautions covering fire prevention practices should be observed in their handling. In addition, the area in which aircraft fuel systems are being maintained should be kept free of flammable materials, oxidizers and acids. Fire hazards are thoroughly covered by National Fire Protection Association (NFPA) standards.

Skin contact with fuels. Fuels may irritate the skin on contact and may cause dermatitis with repeated exposure. Here are some rules those handling fuels should follow:

- Remove liquid fuels from the skin immediately with soap and water.
- Remove any fuel-soaked clothing.
- If the eyes or face come into contact with fuel, flush the area immediately – preferably at an eyewash station. Seek medical treatment promptly.
- If you wear contact lenses, initially flush eyes then remove contact lenses and continue to flush eyes. Before replacing your lenses they should be independently and thoroughly washed.

Confined space entry safety procedures

The aviation setting has the potential for confined spaces for which entry has been regulated by many federal, state and local governments. A confined space is a space that is large enough and configured so that an operator can bodily enter and perform assigned work. It has limited or restricted means for entry or exit, and is not designed for continuous human occupancy. Some of the confined spaces that may be present in the aviation environment include fuel tanks, lateral fuel pits, dikes, sewers, communications and electrical vaults, and heating systems. Safety assessments and procedures must be established prior to entry of any confined space.

Electrical systems and electronic equipment

Short circuits, overloading, accidental grounding, poor electrical contacts, and misuse are all potential sources of accidents involving electrical power in a hangar. All electrical systems used in aircraft – primary power supplies, communications equipment, and electronic equipment – produce or use extremely high voltage. Severe electric shock or electrocution is an ever-present hazard to those repairing electrical equipment, whether it is low-voltage (600 v or less) or high-voltage (more than 600 v). Standard safe practices must be followed for all work on electrical systems.

Use only wooden, fiberglass or plastic ladders when servicing electrical equipment or electrical systems.

Ground fault circuit interruption installations should be provided as established by local electrical codes and as considered necessary.

Personnel throughout the aviation industry, but particularly those in maintenance, are exposed to hazards created by improper use of electrically powered equipment. Serious injury, even death, can result from unsafe use of electrical power. Major damage to materials and property can also result from explosions and fires caused by improper use of electrical equipment.

Inspection. All electrical equipment facilities should be regularly inspected by qualified personnel. Explosion-proof and vapor-tight equipment should be regularly maintained to comply with your local, state or federal authority; or your country's national standards body.

Personnel. Only trained and qualified electricians should be permitted to install and maintain electrical facilities and supply lines. All personnel working on electrical circuits should avoid wearing loose clothing. Also, personnel should not wear jewelry or metal objects that can conduct electricity and thus cause shock, burns or electrocution.

Grounding. Aircraft should be grounded to proper grounding plugs at all times when within hangars or undergoing maintenance. Grounding circuits should be periodically tested.

Ground wires should be located to minimize damage and wear. Large metal equipment complexes, such as aircraft docks, should be permanently bonded and grounded.

Controls. Electrical controls on circuits within hangars should be suitable to the location. They should be positioned to minimize the possibility that anyone will accidentally come into contact with them.

Repair. Only qualified personnel should repair the electrical circuitry in hangars. Electrical circuits to be repaired or installed should have the line switch locked "open" and tagged to prevent the circuit from becoming accidentally energized. Only the person tagging and locking the switch should be permitted to remove the tag and lock.

Warning signs. Adequate warning signs should be posted in areas where hazardous electrical facilities exist, particularly around high voltage apparatus.

Wiring. All wiring within hangars should be installed by qualified electricians, and all conductors protected by conduit or armor. Temporary or makeshift wiring – other than extension lights – should be prohibited in hangars. Operators should assume all conductors are "energized."

Overloading. Electrical circuit overloading is dan-

gerous and should not be permitted at any time. All primary systems within hangars should be equipped with circuit breakers or other accepted means to prevent accidental overloading. Circuit breakers should be reset only after the circuit is tested and the overload has been eliminated.

Cords and receptacles. Electrical cords on portable lamps should be heavily insulated (see National Electrical Code or other appropriate standard). Operators should make sure the cords are long enough for the job, and should avoid stretching and breaking them. Extension cords should not be used. All cords must be inspected regularly for signs of deterioration or damage. Damaged or frayed electrical cords should be replaced immediately. Wall plugs must not be interchangeable with those of different voltages.

Low-voltage systems. Low-voltage systems (600 v or lower) should receive the same respect given high voltage circuits. Severe shock and death can be caused by 100 v or less if a good contact is made. Regardless of voltages, follow safe practices.

High-voltage systems. High-voltage systems (higher than 600 v) expose all service people to hazards. The treachery of electrical voltage is that its presence cannot be detected by human senses until injury occurs. Therefore, everyone should treat all electronic and communication devices with caution.

Standard practices should require power to be turned off for work on electronic, communication or energizing equipment. If an adjustment can be carried out only when the circuit is "hot," or energized, the servicing should be done by qualified personnel who are properly trained.

Repair operators should not use metal rules or uninsulated tools near electrical voltage circuits. Service personnel should not use hand tools (such as screwdrivers) to make energy tests on any piece of electronic equipment. Serious burns can result from this practice.

Circuits should be de-energized when work is in

progress. Equipment should be grounded according to manufacturer's instructions and procedures. Grounding "hot" circuits should be done at the exact point of the intended repair or adjustment prior to making contact with that particular area of the energized equipment. (Grounding may be needed on both sides of the circuit or equipment). Grounding should be used on all circuits being worked on that could contain residual energy, such as capacitors, or circuits that could be energized. It should not be removed until repairs are completed. The system should not be energized until grounding is removed.

Equipment safety devices, such as interlock switches, disconnect relays, or automatic circuit breakers are designed to give protection to electrical circuits and equipment. Because these devices are subject to failure, repair operators should take other necessary precautions, such as manually disconnecting or locking out control switches when working on electrical equipment. Safety interlocks should not be deliberately decommissioned. Where appropriate, nonmetallic ladders and extension cords with ground fault circuit interrupters should be used.

When antenna repairs are undertaken, power should be grounded out at the transmitter, and a warning tag indicating work in progress applied.

Electronic equipment carrying high voltages (higher than 600 v) should not be unplugged with the power on.

High voltage switches. Electrically operated high voltage switch gears should not be manually closed on any energized electrical circuit.

Radar systems. Neither weather radar nor any other type of radar emission should be allowed in hangars because radar in sufficient intensity may cause damage to human tissue, particularly to the eyes. Another reason to exclude radar operations from hangars is that radar beams might activate hangar warning systems by causing localized heating. Radar is also dangerous near fueling operations for the same reason.

Cathode-ray tubes. Because of the high vacuum present in cathode-ray tubes (CRTs), anyone working on them should prevent bumping, scratching, or cracking them. A violent implosion may result if a CRT is damaged or handled too roughly. In the event of an equipment fire in a chassis that has CRTs, use of carbon dioxide (CO_2) gas extinguishing agents should be avoided. As the gas expands, it becomes very cold. The contact of the cold gas with the warm glass of a CRT will result in a violent implosion. Halon extinguishing systems are preferred.

When maintaining or removing CRTs, operators should wear safety glasses. They should also wear rubber gloves as protection against potential implosion. Proper grounding procedures should be followed.

Engine starters. Like other electrical apparatus, high energy ignition systems have the potential to kill. Therefore, only employees who are properly trained and technically qualified should work on high energy ignitions.

Control systems in aircraft

Serious damage can be done to aircraft (and personal injury may also result) if aircraft controls are operated without the use of correct procedures. Here are some basic procedures for operating controls:

- Mechanics should make certain that all personnel and equipment are clear of the moving surfaces on the aircraft before operating any controls in the flightdeck. Use radio communications from ground to flightdeck when tests are carried out.
- Place placards on the controls in the flightdeck with warning signs when work is in progress on either surfaces or linkages. Only maintenance personnel actually involved in doing the work should remove the placards.

Walking and working surfaces

In the United States, OSHA regulation 1910.28 specifies that walkways, scaffolds and stairways used to service wing and tail assemblies on aircraft should be constructed of materials sufficiently strong to provide a safety factor of at least 4. (A safety factor is the ratio of the ultimate [breaking] strength of a member or piece of material to the actual working stress or to the maximum permissible [safe load] stress when in use.) Walkways, scaffolds, and stairways should be of stable design and adequately secured to the floor or aircraft dock. Work platforms more than 1.2 m (4 ft) in height and 0.75 m (30 in) above the floor or surround area should be equipped with guardrails and more than 3.2 m (10 ft) should also be equipped with toeboards, in compliance with your local, state, or federal authority or your country's national standards body. Additionally, toeboards may be required at lower heights wherever, beneath the open sides persons can pass; there is moving machinery; or, there is equipment with which falling materials could create a hazard. (Figure 18-1).

When the nature of the repair makes it necessary for the operator to step on the aircraft, the operator should use designated walkways. Under no circumstances should personnel step on control surfaces, air scoops, fairings or other parts that are not designed for walking. All walkways, scaffolds, and platforms should be suitably guarded for the work to be undertaken. Personnel working on unguarded, elevated surfaces should wear lifelines and safety harnesses.

Horseplay in any form should be forbidden in and around all aircraft during maintenance operations. Horseplay is dangerous to both people and aircraft.

Aircraft pressurization and air-conditioning systems

Air control systems. These systems on an aircraft are usually composed of a turbocompressor assembly, heat extraction assembly, reheat assembly, and valves and ducts. Maintenance on air control systems inside a hangar should be confined to functional checks of components. Pneumatic test equipment, which involves ducting to carry high-temperature compressed air to an aircraft, is widely used to test pressurization, air conditioning and pneumatic installations on an aircraft. From a personal safety point of view, external lines carrying high-temperature compressed air to an aircraft should be

Figure 18-1. Aircraft service docks should be strongly built of non combustible materials. Operators at elevated heights should wear safety harnesses and lifelines.

Photo courtesy of Northwest Airlines

well-insulated and installed so as to be inaccessible to operators. Air ducts and nozzle attachments should be maintained in the best condition possible and should have locking devices built into an aircraft coupling. Standard fan guards with mesh small enough to prevent fingers from contacting fan blades must be provided.

Turbocompressor run-in. Temperatures and pressures of intake air should not exceed published limits. Guards for compressors should be installed and kept tight during run-ins. No differential pressure should be permitted to build up in the cabin of an aircraft unless this is a requirement of the test and well-posted.

Freon compressors and condensers. Compressor guards should be in place and tight during functional tests or run-ins.

Heaters. Where heaters are an integral part of the air-conditioning system, their exhaust should go into the atmosphere during tests.

Spill valve test. The pressure surge necessary to trip the spill valve should not exceed maximum published pressure.

Anti-icing functional test. For such tests, the control surfaces may be heated to as much as 200° F (93° C). Personnel in the area should be warned of all hazardous hot surfaces.

Pneumatic pressure. This is used inside hangars to actuate certain controls of the aircraft, such as the reverser mechanisms of certain turbojet engines. Aircraft flightdeck controls and selectors are to be posted with warning placards when pneumatic pressure is being supplied to an aircraft. When controls are being tested pneumatically, no one should be in a position to be injured and all personnel in the area should be made aware of the test.

Landing gear

Landing gear maintenance involves the use of aircraft jacks and hydraulic lift tables. To protect personnel from injury, landing gear and the elevating means must be carefully monitored, and prescribed procedures strictly followed.

Aircraft jacks. These are the basic principles to jacking operations:

- All jacks should be properly rated for the job for which they are being used. They should be in good condition and routinely inspected.
- Place jacks firmly on a flat surface. Correctly position jacks under the aircraft jackpoints, as identified by the manufacturer.
- Alert all personnel in the area of the jacking operation.
- Make sure that the area around the aircraft, on all sides, is clear, in case one or more jacks fail under the load or the aircraft moves off the jackpoints.
- After the aircraft is jacked to the desired height,

make sure that the jacking controls are protected from accidental operation.

Hydraulic lift tables. Follow these procedures before the aircraft is positioned:

- Check the hydraulic lift tables to see that their controls, wiring surfaces, and fluid levels are in order.
- Check that the table operates full travel.

Follow these procedures for positioning aircraft:

- Alert all personnel that an aircraft is being positioned for jacking.
- Check that the jacks are properly placed and serviceable.
- Position the aircraft on the table in line with the jack assembly.
- Lower both tables simultaneously with the rear wheel assembly. The tables should stop just clear of the jackhead.
- Check the alignment and the lower tables until the weight is off the undercarriage.
- Lower the aircraft onto the nose jack.

Undercarriage work

This section covers basic procedures for work on the undercarriage of aircraft.

For removal of wheels or an assembly:

- Use a table to support the part being removed.
- When the component is disengaged from the wheel assembly or the aircraft, use the necessary equipment to roll it away.

For a swing test of undercarriage assemblies:

- Lower the table or raise the jacks until all the wheels are clear of the floor.
- Alert all personnel of the test.
- Check that no mechanics are working in the wheel-well area of the aircraft.
- Begin the test only when the area is clear, roped off and posted with appropriate signs.

After a swing test:

- Warn all operators before the aircraft is lowered.
- Lower the aircraft by degrees to make sure that the gear can take the full weight of the aircraft.

It is recommended that tires be inflated only when enclosed in a tire cage. They should be inflated in workshops whenever practical. When it is necessary to inflate tires in hangars, use appropriate metering equipment for the job. Under no circumstances should personnel inflate tires or struts directly from a high-pressure cylinder.

Radiography

Radiography in the aircraft industry involves:

- Radioactive isotope iridium-192 (Ir-192)
- Portable X-ray units of 130 to 200 KVP
- Transport and storage facilities for radioactive materials

All procedures concerning radiography must be in accordance with appropriate local, state and federal regulations.

Iridium-192. This radioactive isotope should be stored in a lead-lined room, radiation-free to 165 KVP in any direction. The isotope itself must be encapsulated at all times, with only beams exposed when required. Weekly readings must be made of the lead-lined room, and readings must be taken of the capsule before every use.

X-ray units. These units should be stored and locked in isolated confines when not in use. Daily readings should be taken of the shielding of the x-ray tube head to trace any radiation leakage through the tube walls.

Personnel safety precautions. Technical personnel working in radiography should at all times wear two pocket dosimeters. These should be zeroed every morning and checked at the end of each shift. To determine cumulative dosages, exact dosimeter reading records should be maintained for each person. In addition, each operator should wear a film badge, and the results should be checked by an out-

side agency. Permissible dosage limits are available from government regulatory bodies.

Each employee working with radiation should receive a preplacement medical examination. A blood count should be taken of each employee every six months; a full annual medical checkup is also recommended.

When X-ray equipment is set up in a workshop or hangar, all employees not associated with the operation should be cleared from the area. Using meters, the X-ray technicians and operators should establish a safe boundary limit and mark it off with standard warning signs. In addition, audible or flashing light warnings should be used to alert personnel of the operation being conducted.

The area must be constantly surveyed during operation of X-ray units. The location should be checked for scattered radiation patterns. Each time the location of use is changed, the new area must also be checked for scatter patterns

WORKSHOP OPERATIONS

Careful adherence to accepted safety standards and precautions will greatly reduce the hazards present in workshop areas.

The principles in this section should be taught to, and practiced by, workshop personnel in their regularly assigned jobs. Supervisors, ground safety personnel, and others responsible for training and safety should familiarize themselves with these standards. They should assume the responsibility for instructing subordinates in the various provisions.

The medical department, which is responsible for evaluating operators' physical capability for the job requirements, should recommend appropriate action when examinations indicate that operators do not have physical qualifications for a particular job.

General principles

Principles to be followed in the physical aspects of a safety program for workshop operations are:

Analyze. Analyze the potential hazards of each job and provide safe methods, materials, procedures, facilities or equipment.

Isolate. Operations that may be potentially hazardous to passers-by or people not specially equipped should be isolated by a screen or done in a separate room, if possible. Painting, for example, must be done in a separate, well-ventilated room. Small welding jobs may only have to be screened off.

Guard. Provide a barrier guard that prevents the operator from putting hands or any part of the body in danger.

Layout. For safe, efficient production, be sure operations and facilities have proper lighting, ventilation, aisle and work space.

Identify. Use colored lights, barriers, signs, audible signals and other means to mark potentially hazardous areas or operations.

Principles of machine guarding

Proper machine design and installation of guarding removes much of the threat to safety for the operator of a machine. There are two general classifications of machine guarding:

1. Power-transmission guarding (including all equipment from the prime mover to, but not including, the point of operation)

2. Point-of-operation guarding (where the actual work of the machine takes place)

In the interest of safety and economy, machinery should be purchased with appropriate guarding. Establish a procedure that the purchasing staff consults the safety department on current safety standards or specific needs for optimum protection of machinery before buying equipment.

Guards should be designed to give maximum protection to the operator yet not interfere with the normal operation of the equipment. For instance, in

order to permit changing drive belts, making adjustments, or lubricating, machine guards may have to be designed with hinged or removable sections.

When machinery and power-transmission equipment do not have guards as part of their design, suitable enclosures or barricades should be installed to prevent injury resulting from persons contacting moving parts or hazardous substances.

Machine shop layout

For maximum safety, machines should be located to provide sufficient space for the operator to handle materials and perform routine job operations without interference from his or her own equipment or nearby operators. Machines that may move or topple over during operation should be firmly secured to the floors, bases or stands.

If equipment contains heavy or high-speed moving parts that may disengage or disintegrate and strike nearby persons, it should be located behind shielding that will fully contain flying fragments.

Lighting at the point of operation and in the immediate areas should be adequate for performing tasks and designed to eliminate excessive glare or shadows.

Materials-handling equipment should be installed to assist the operator in handling awkward or heavy parts.

Machine controls should be conveniently located for the operator. Stop switches or power disconnects should be located within easy reach of operators. Controls for drag lines, conveyor systems, and like equipment should be installed to allow for immediate emergency shutdown.

At no time should stop switches be made inaccessible by covering them or blocking them off. Power controls or disconnects should be locked in the "off" position when repairs are necessary.

Machines, or their appropriate parts, should be color-coded according to established standards to help distinguish them and to improve visibility, thereby reducing the accident potential.

Handling hazardous materials

All hazardous materials used in workshops should be subject to appropriate, effective control measures.

The most effective means of controlling hazardous materials is to substitute a nonhazardous material for a harmful type. Where a substitution can be made, make certain that a different hazard is not created by the new material.

The procedures to follow for the control of hazardous materials include these:

- Inventory all chemicals and other materials with hazardous properties used in the workplace. Acquire the appropriate Material Safety Data Sheet (MSDS) for the materials. These sheets list the hazards of a substance, give precautions on safe handling, and recommend control measures.
- For hazardous materials received without adequate labeling, properly identify and label them. Review government regulations on process or transfer container labeling requirements. Depending on your government regulations, it may be necessary to report chemicals manufactured by your organization and sold to others.
- Inform operators who may be exposed to hazardous materials. Operators should know the hazardous substances to which they are exposed, their nature, the protective measures to take in handling them, and the safe practices required on the job. Maintain a record of such training.

NOTE: MOST COUNTRIES HAVE A REQUIREMENT FOR MANAGEMENT TO ADVISE THEIR EMPLOYEES OF WHAT SUBSTANCES THEY ARE WORKING WITH AND THE REQUIRED PRECAUTIONS TO BE TAKEN TO ENSURE A SAFE WORKING ENVIRONMENT.

- Establish a means to notify contractors in your facility of the hazards the contractors' employees may be exposed to while working on the facility.
- Establish a written policy and training program on

hazardous materials to conform to legal requirements.
- Describe emergency procedures and the nature of the hazard in the program.

Personal cleanliness is most essential in operations involving possible skin contaminants. Operators should be informed of the need for personal cleanliness. In order to encourage and facilitate personal cleanliness, suitable washrooms, showers, and locker rooms should be available.

Supervisors should make certain that all safety regulations and standards are strictly observed. They should give short safety talks at frequent periods to all employees under their jurisdiction.

Particular hazards in workshops

Fire hazards. In areas where flame-producing equipment is used, safety controls should be established. Smoking should be forbidden except in designated areas.

Workshops should be adequately ventilated to prevent the accumulation of hazardous gases and vapors. Special exhaust systems should be provided in battery rooms, painting booths, and for welding jobs.

Carbon monoxide. Carbon monoxide is one of the common gases to which personnel in vehicle workshop operations are exposed. Flexible tubing attached to vehicle exhaust pipes should be used to carry carbon monoxide and other exhaust gases outside the workshop. General exhaust ventilation should also be provided to prevent accumulations of this poisonous gas.

Floor drains and spills. Shop floor drains should not be connected to sanitary sewer lines. Whenever possible, workshop operations should be laid out so that flammable liquids will not drain into floor drain systems. Such liquids should be drained into suitable containers. The containers should be held for disposal by a reputable treatment and disposal facility.

Emergency procedures for material spills as recommended by the manufacturer (refer to MSDS) should be followed in the event of a spill. If the spill enters a public sewer system the local agency should be contacted.

Boxes of suitable absorbent materials should be provided in vehicle maintenance workshops to be used on grease or oil spills. After absorbent material has been applied to a spill, the floor should be thoroughly cleaned.

Workshop floors should be constructed to slope toward drains equipped with grease traps. These traps should be cleaned out at frequent intervals.

Welding hazards. Welders should wear the necessary gloves, aprons, helmets, or shields when actually performing welding operations. They should also be careful to wear filter goggles of the shade appropriate for the type of welding equipment being used. Before welding or other heat-producing work is performed on a vehicle, gas tanks and other fuel containers should be removed from the vehicle. They should then be drained, purged, and filled with water or otherwise rendered safe by filling them with inert gas.

Battery handling. Workshop personnel handling batteries should be provided with rubber gloves, aprons, and suitable eye protection. When handling electrolyte, operators should wear chemical safety goggles and face shields, rubber gloves, and aprons. A water source suitable for the size of the operation should be provided for operators to wash off acid that may splash on their eyes or bodies. There should be a continuous flow eyewash station in the immediate vicinity.

Machine shop operations. Maintenance personnel using grinding wheels or cutting tools that produce flying chips or dust should wear impact safety goggles or face shields. Mechanics working under vehicles should wear safety goggles to prevent dirt from entering their eyes.

Workshop machinery, such as lathes, abrasive wheels and portable electric tools should be guarded according to federal, state or provincial standards.

Pits. Guardrails should be placed around grease or repair pits. All pits should have steps and handrails to permit safe entrance and exit. Pits should be built with drains equipped with oil or grease traps. Pit drains should not be connected to sanitary sewers. In addition, there should be a local exhaust system to remove flammable vapors.

Repair and grease pits should be cleaned regularly with soap or cleaning solution and water or a caustic solution. Pit contamination is a potential health, fire, and accident hazard. In order to prevent accumulation of hazardous gases, an engine normally should not be operated while a vehicle stands over a repair pit unless exhaust gases are safety ducted to the outside, away from fresh air intakes.

Compressed air. Compressed air is used in many workshop operations, such as spray painting, tire inflation, and fuel line cleaning. When handled properly, compressed air is not hazardous. However, many mechanics are tempted to use air under high pressure to clean work benches, vehicle chassis, parts, or even their skin or clothing. This practice is extremely dangerous because the air pressure blows around metal parts, dirt and debris.

If the air pressure is strong enough, flying particles can be forced through clothing, or into the flesh or eyes. Air itself can be forced into the bloodstream, with potentially fatal results. Mechanics never should engage in horseplay with compressed air.

Compressed air pressure should be reduced to minimal operational requirements and should never exceed 207 kPa (30 psi) for cleaning purposes. Chip guarding should be provided. Approved nozzles are to be used on air hoses.

Controlling hazards. Whenever practical, processes creating health or safety hazards should be isolated as completely as possible. Physical isolation in circumstances where personnel must work nearby may mean the use of protective shields, barriers, and other similar devices. Noise-attenuating or noise-deflecting barriers may be advisable.

Certain operations may call for employees to wear personal protective equipment such as respirators, hoods and helmets, safety glasses, face shields, hearing protection, and gloves (Figure 18-2). Rules for the use of such equipment should be enforced.

Ground equipment maintenance

A procedure should be established to ensure that ground service vehicles are checked daily. Units that an operator finds to have defective components that affect the safe operation of the vehicle should be removed from service immediately, tagged with a description of the defect or defects, and routed to the workshop for correction before being returned to service.

Inspection. Inspections of ground vehicles should cover these items:

- **Brakes.** Check floor brakes for smooth and effective operation. When the brakes pull a vehicle to one side or the other, there is a braking deficiency. Such deficiencies should be corrected immediately. Hand brakes should be capable of holding vehicles on a reasonable grade when only two-thirds of the handle travel has been taken up. (Brake pedals and clutch pedals, when applicable, should have a skid-resistant surface.)
- **Lights.** Check all lights on a vehicle for burned-out lamps. Adjust headlight beams to ensure the area in front of a vehicle is properly and evenly illuminated, without glare to approaching drivers. Operate switches and trace any loose or broken connections to their source and repair. Check that the lenses and reflectors of vehicle lights are clean and securely fastened in place. Check directional signals and backup lights, if applicable.
- **Windows.** Check that the windows of all vehicles are clean and free of obstructions. Cracked windshields and windows will distort and limit a driver's vision and endanger the safety of drivers

Figure 18-2. Each successful project is a result of a combination of factors, including proper equipment and materials, skillful operators, good procedures and correct protective gear.

Photo courtesy of Air Canada

and protection of property. They should be replaced as soon as possible.
- **Rear-view mirrors.** Every vehicle should have at least one rear-view mirror. Check that these mirrors are clean and properly adjusted. The mirror fastening should be secure enough to prevent dislocation by vibration while the vehicle is moving.
- **Horns.** Test vehicle horns for satisfactory operations and sufficiently loud warning signals.
- **Windshield wipers.** Check windshield wipers for satisfactory operation. Check that blades and arms are in place and securely connected. Blades must contact the glass properly to provide a driver with a clear field of vision.
- **Seats.** If removable seats are installed, the clamps should hold them securely.
- **Fire extinguishers. For vehicles equipped with fire extinguishers,** check that extinguishers are mounted in an accessible place, readily removable and protected from damage. They should be inspected, serviced and ready for use at all times.
- **Towing connections.** Towing connections should be maintained in good condition. Check

the locking mechanisms for security. Operators should connect the brake lines and safety chains before moving the tow vehicle-trailer, when applicable.

- **Wheels.** Before operating vehicles, drivers should check the wheel mountings, rims and axle flange nuts to be certain they are secure. Tires should be gauged and inflated to specified pressures at the frequency recommended by the manufacturer.
- **Other items.** Passenger stands, ladders, work stands, steps and handrails should be checked for good repair and security. Skid-resistant material should be replaced if worn or missing. If equipped with towing hitches, they should be inspected for visual cracks or worn mechanisms.

Fueling of vehicles. Vehicles should not be fueled indoors as a routine practice. Indoor fueling may sometimes be performed under controlled conditions – and then only when approved by the fire marshal or other authority having jurisdiction.

Vehicles taking on fuel, and those not in use and unattended, should have their engine stopped and the parking brake applied. No vehicle should be operated unless entirely free of fuel leaks. During fueling, the nozzle of the hose should constantly contact the intake pipe of the vehicle's fuel tank to provide an effective bond and prevent the discharge of static sparks.

Maintenance operations. Motor vehicle maintenance activities such as painting, welding and battery work should each be carried on in a separate part of the workshop where the operations of one kind will not be hazardous to another. Adequate general illumination should be provided for all operations, and – where necessary – include supplemental lighting.

Vehicle lift safety
All vehicle lifts should be equipped with a safety device to prevent accidental lowering. The device can be a simple mechanism such as a safety leg, which locks in a vertical position as the lift is raised; or it can be a restricted orifice device, which permits controlled lowering in the event of a hydraulic failure. As an added safety feature, air-oil-operated hydraulic lifts should be equipped with a lock that prevents raising by air if the oil supply is low. A safety device meeting this requirement is called a "low-oil lock." This is a removable device that prevents compressed air from entering the lift cylinder assembly; the arrangement prevents the plungers from being raised above the oil supply in the air-oil reservoir.

Roll-on lifts should be equipped with stop chocks to prevent the vehicle from moving while the lift is hoisting, lowering, or in the elevated position. Preferably, stop chocks will automatically spring into position when the vehicle enters the lift.

Drivers should not be permitted to remain inside vehicles that are on moving or elevated lifts, unless a specific operation requires it. If on an elevated lift inside a vehicle, the driver must keep the doors closed to reduce the fall potential.

All automotive lifts should be marked with the name of the manufacturer, lift capacity, and date of installation. These markings should be stamped or etched on a metal plate permanently attached to the lift in a position where it can be checked.

All vehicle lifts should be equipped with "dead-man" controls, which automatically turn to "neutral" or "off" when released by an operator. Controls should be located near the lift.

Chassis and axle supports should be designed to transfer the load to the lift rails. No makeshift devices should be used for chassis and axle supports.

Other maintenance precautions
Workshop employees should use mechanical devices, such as dollies or hoists, to help them mount or remove large, heavy tires.

The rated load of jacks should be legibly and permanently marked on them in a prominent location by casting, stamping, or another suitable means. Operators should make sure that the jacks have a rating sufficient to lift and sustain the load. Jacks should also be inspected and tested in compliance with your local, state or federal authority; or your country's national standards body.

After a vehicle has been raised by a jack, it should be securely blocked to prevent falling. At no time should mechanics place any part of their bodies directly under the wheels of a jacked or blocked vehicle.

Mechanics should keep their clothing free of grease and oil. Clothing contaminated with flammable substances should not be worn or stored in clothing lockers. Operators should not wear neckties, loose clothing, wristwatches, rings, or other jewelry when working near or with equipment.

All hand and portable power tools should be used and maintained according to the highest standards. Tool kits should be inspected periodically and defective tools replaced immediately. For detailed information on tool inspection and maintenance, see the National Safety Council's *Accident Prevention Manual for Industrial Operations – Engineering and Technology*. (See References.)

Power cables and cords on portable electrical equipment should be made of heavy, armored rubber, or similar material, to prevent damage from oil, grease and chafing. The cables should be constructed to provide automatic grounding of the equipment through integral conductors. Mechanics using portable electrical tools and lights should not string cables across workshop floors. Serious injuries can be caused by tripping over power cords and cables.

Workshop traffic

Workshop entrances and exits should be clearly marked and lighted to prevent accidents. Appropriate traffic signs should be posted at entrances and exits. Drivers entering or leaving workshops should sound their vehicle's horn to warn personnel of oncoming traffic. In the interest of safety, a maximum limit of 8 km/h (5 mph) should be enforced in and around workshops.

Driveways and the areas around workshop entrances should be kept clear of ice and snow and treated with

sand, ashes, calcium chloride, rock salt or other similar materials if these areas become slippery. Workshop floors and inside entrances should be cleared of ice and snow brought in on the wheels of vehicles.

Paint shop

Painting operations present three potential hazards to health and safety: fire, toxicity and mechanical hazards.

Fire. Flammable paints and thinners must be kept away from sources of ignition. They should be stored in approved flammable liquids storage cabinets. Housekeeping standards should be meticulously maintained. Masking materials, paper and soiled rags used in painting should be deposited in self-closing metal containers. Combustible materials stored in hangars and workshops should be kept to a minimum. Paint booths must be posted with "no smoking" signs. All paint-dispensing containers and drums must be properly grounded to ensure against static electricity buildup.

Toxicity. Painting is to be done in approved paint booths with mechanical ventilation. Employees who paint outside of booths should wear properly fitting, approved paint respirators at all times and should change filters regularly.

NOTE: ALL EMPLOYEES REQUIRED TO WEAR RESPIRATORS SHOULD BE TRAINED IN THEIR PROPER FIT, TESTING FOR FIT, USE, CLEANING REQUIREMENTS AND LIMITATIONS.

Protective clothing made of material other than rayon, nylon, silk, wool and certain plastics should be worn by operators who paint. For optimum protection against static electricity buildup from protective clothing, wear cotton.

Mechanical hazards. Spray-painting equipment should be kept clean and in good repair. Pressure hoses should be inspected for wear and replaced as necessary. Care should be taken in placing pressure hoses so that people will not trip over them and lines will not be cut by passing vehicles. Workstands should be of the correct height and size so painters can work with reasonable ease and comfort. High-pressure paint nozzles should not be pointed at any operator. The recommendations contained in NFPA 33 (see References) should be closely followed for these operations.

Battery charging

Potential hazards in battery shops are corrosive or toxic chemicals, and explosive gases. Battery shop operators should be thoroughly trained in their jobs and familiar with safe operating procedures. Nickel-cadmium and silver-zinc batteries should be serviced in an area isolated from lead-acid batteries. Sufficient explosion-proof ventilation should be provided to prevent acid fumes from entering the nickel-cadmium or silver-zinc battery shop. When both acid and potassium hydroxide electrolyte batteries are handled in the same shop, the equipment for the two kinds of batteries should be kept separate and carefully labeled. Implements that have been used to service nickel-cadmium alkaline, silver-zinc, or nickel-iron alkaline batteries should be used on those types only.

Tools or metal parts should not be laid on a battery or stored in such a position that they may fall on a battery. Operators should not wear rings, wristwatches, or other jewelry. They should use nonsparking lifting devices and explosion-proof (nonsparking) hand trucks to handle heavy batteries. When manual lifting is necessary, operators should obtain sufficient help to prevent injury.

Lead-acid battery charging operations should be located in well-ventilated, rooms. Excessive charging of lead-acid batteries should not be permitted, because this produces excessive quantities of hydrogen gas, creating a highly explosive gas mixture in the shop's atmosphere. A lead-acid battery should be allowed to cool down in accord with specifications after removing it from the charger. Work should not be done immediately on a battery that either has run down under heavy load or has just been charged. In both cases, the battery may be gassing rapidly and may explode if heat is applied.

Silver-zinc batteries do not generate harmful gases during normal charging or discharging, so there is no need for exhaust fans and nonsparking equipment.

To prevent burns when handling electrolyte, shop operators should wear protective chemical safety goggles and face shields, rubber gloves, and aprons. When mixing electrolyte, they should always pour the acid or potassium hydroxide into the water slowly. The water should never be poured into the chemical because rapid heat generation may cause explosive steam expansion. A tilter should be used to pour acid from a carboy. When acid is siphoned, operators should use a rubber suction ball – never their mouths – to start the action. Deluge showers and continuous flow eyewash fountains must be provided in all battery shop operations.

Smoking and open flames should be strictly forbidden in battery shops. Care should be taken to remove unnecessary sources of ignition, including electrical-spark-producing devices, from the immediate area of battery vents when hydrogen concentrations are possible. Electrical facilities and spark-producing equipment used in battery shops should conform to the explosion-proof provisions of your local, state, or federal authority or the national standards body of your country, such as the National Electrical Code (see References).

Batteries should be handled with extreme care to avoid spark-producing short circuits. Operators should use terminal straps to handle batteries at all times. These straps should be kept free of grease or other foreign substances.

Floors in the battery shop should be kept clean and dry to prevent injuries from slipping and falling. Floors and battery racks should be made of acid-proof materials like finished oak. Floors should be covered with acid-resistant paint. If metallic battery racks are used, they must be grounded to prevent sparking. (See CFR 1910.178(g).)

Tire service

Workshop personnel should use mechanical devices, such as dollies, to help them mount or remove large, heavy tires.

Guard cages should be used when inflating tires, particularly those mounted on wheels with lock rings. When such wheels are mounted on vehicles, several lengths of chain, fitted with positive catches, should be passed through the wheel openings and around the tire rim before inflating. Inflation should only be done with a "clip-on" airchuck. A rim lock ring that releases from a wheel during tire inflation or mounting can be propelled from the tire at a lethal velocity (Figure 18-3).

MATERIALS HANDLING

Operators should receive thorough instruction in safe lifting and carrying procedures, as well as in the use of personal protective equipment for handling materials. Adequate supervision is necessary to ensure that safe practices are being observed at all times by those who handle material. Wearing rings, jewelry, loose clothing, and improper footwear during handling operations is extremely hazardous and should not be permitted.

Proper planning in materials handling is also an important safety element. Materials should be handled in logical sequence, and unnecessary handling

Figure 18-3. A worker prepares to inflate a tire, using a remote chuck.

Photo courtesy of Piedmont Airlines

procedures should be avoided. Mechanical lifting devices should be employed where ever possible.

The principles in this section apply in all areas – including workshops and hangars.

Manual handling

These factors must be considered in establishing safe lifting limits for operators: physical differences, periods of sustained lifting, height of lifting, distance and frequency that the load is carried, and the size and shape of a load. (For more specific information, see your local, state or federal authority; or your country's national standards body. See also References, NIOSH Work Practices for Manual Lifting and Lifting Calculator.) Operators should be required to take a medical examination to determine the suitability of assigning them to manual handling duties.

General pointers that can be given to those who handle materials include:

- Inspect materials for slivers, jagged or sharp edges, burrs, and rough or slippery surfaces.
- Grasp the object with a firm grip.
- Keep hands and fingers away from pinch and shear points, especially when setting down materials.
- When handling long objects, keep hands away from the ends to prevent them from being pinched.
- Wipe off greasy, wet, slippery or dirty objects before trying to handle them.
- Keep hands free of oil and grease.

In most cases, gloves, hand leathers or other hand protectors should be used to prevent hand injuries. Extra caution must be used when working near moving or revolving machinery.

Use of body weight. Employing the correct positioning of the feet and the flexion and extension of the knees, the weight of the body can be effectively used to push and pull objects and to initiate a forward movement, such as placing an object on a shelf or walking.

When lifting an object from the ground, the thrust from the back foot, combined with the extension of the knee joints, will move the body forward and upward, and for a brief period it will be off balance. However, this position is immediately countered by bringing the back leg forward, as in walking, and by this time the lift is completed. This forward movement of the body results in a smooth transition from lifting to carrying. Do not "jerk-lift" a load as this multiplies the stress to the lower back.

Here are some techniques for specific situations.

- If the object is too bulky or too heavy to be handled by one person, help should be sought.
- Before lifting the load to be carried, the operator should consider the distance to be traveled and the length of time that the grip will have to be maintained. The operator should select a place to set the load down and rest in order to allow for the loss of gripping power during a long-distance carry. Pausing to rest is especially important when negotiating stairs and ramps.
- To place an object on a bench or table, the operator should first set it on the edge and push it far enough onto the support to be sure it will not fall. The object should be released gradually as it is set down. It should be moved into place by pushing with the hands and body from in front of the object. This method prevents pinched fingers.

It is especially important that an object, placed on a bench or other support, be securely positioned so that it will not fall, tip over, or roll off. Supports should be correctly placed and strong enough to carry the load. Heavy objects should be stored at approximately waist height.

- To raise an object above shoulder height, the operator should lift it first to waist height, then rest the edge of the object on a ledge, stand, or hip, and shift hand position, so the object can be boosted after the knees are bent. The knees should be straightened out as the object is lifted or shifted to the shoulders. As in the case of lifts originating at ground level, the upper extreme of muscle stress begins at shoulder height. Good job design should be applied to eliminate or minimize above shoulder lifting.

- To change direction, the operator should lift the object to the carrying position and turn the entire body, including the feet. Avoid twisting the body. In repetitive work, the person and the material should both be positioned so that twisting the body when moving the material is unnecessary.
- To deposit an object manually in a tight space, it is safest to slide it into place with the hands in the clear.

Instructions for three major lifting techniques are given below.

Squat lifting techniques. Here are six steps to good squat lifting techniques:

- Keep feet parted – one alongside, one behind the object.
- Keep back straight.
- Tuck elbows and arms in, and hold load close to body.
- Grasp the object with the whole hand.
- Tuck your chin in.
- Keep body weight directly over feet.

Assisted one-hand lift. In this method, the operator rests one hand on top of the container, bends over to grasp an object in the container, and then pushes down with the nonlifting hand resting on top of the container to force the upper body back to a vertical position. This method features using the large arm and shoulder muscles to perform the lift instead of the vulnerable muscles of the lower back. By using the nonlifting hand to raise the upper body, stress is distributed across the shoulders and arm muscles and reduced in the lower back.

For a good assisted one-hand lift, follow these steps:

- Place the nonlifting hand on the container top, bend over container and assume lift position.
- While bending over, kick the foot on the same side as the nonlifting hand rearward to provide forward body balance (optional).
- Reach and grasp object to be lifted.
- Push down with the nonlifting hand on the

container top, raising the upper body to a vertical position. (Be sure to let the nonlifting hand, not the back, do the work!)

As with any lifting technique, there are limitations to its use. The application of the assisted one-hand lift assumes that the object to be lifted can be grasped by one hand, is not so long that it becomes awkward to handle and is not of excessive weight.

The assisted one-hand lift technique also may be used to lift objects off the floor or pallet flats by placing the nonlifting hand on a spot above the knee and pushing down on it after grabbing an object with the other hand.

Team lifting and carrying. When two or more people carry one object, they should adjust the load so that it rides level and so that each person carries an equal part of the load. Test lifts should be made before proceeding.

When two people carry long objects, they should carry them on the same shoulder or walk in step. Shoulder pads will prevent cutting of the shoulders and reduce fatigue.

When a gang of operators carries a heavy object, the supervisor should make sure that proper tools are used. Work direction should be provided; frequently, a whistle or direct command can signal "lift," "walk" and "set down." The key to safe gang carrying is to make every movement in unison.

Mechanical handling: basic principles

Mechanical materials handling equipment should be used whenever loads are too heavy or bulky to be lifted and carried safely by hand. Forklifts, conveyors, hand trucks, chutes, rollers and cranes, when properly used, simplify materials handling and greatly reduce hazards.

Although mechanical handling equipment is intended to reduce hazards involved in manual handling, unsafe use of hoists and trucks presents increased danger.

Only fully trained or adequately supervised operators should use such equipment. They should be put through selection and training programs similar to those required for motor vehicle operators. Each operator should be given complete instructions in the proper use of equipment, and the training should include safe standards of operation, inspection, preventive maintenance and materials-handling practices.

Where mechanical handling equipment is used, the aisles should have sufficient clearances for the equipment. There should be adequate clearances at loading docks, through doorways, and where turns must be made. Aisles should be at least 2 ft wider than the widest vehicle in use, and they should be clearly marked by guidelines painted on the floor. Obstructions, such as columns and posts, should be painted, preferably with diagonal stripes of black and yellow. Wooden guards, preferably painted with black and yellow stripes, should be used at the corners of stacked materials to prevent accidental dislodging and to protect the stacks.

Where handling equipment run by internal combustion engines is used in closed areas, there should be adequate ventilation to prevent accumulations of dangerous carbon monoxide gas. If natural ventilation is not sufficient, or forced ventilation cannot be provided, the engines should be equipped with enclosed exhaust piping direct to the outside away from outside air intakes. The piping system should be checked for leaks. As an alternative, electric-powered machines might be used.

Hand trucks should be pushed rather than pulled whenever practical. Truck handles that expose operators' hands to possible injury should be equipped with knuckle guards. When trucks are not in use, the handles should be locked in a vertical position. Positive-action braking mechanisms should be provided on all hand trucks.

Loading handling equipment

On all handling equipment, the center of gravity of the load should be kept low by placing the heavier objects on the bottom and the lighter objects on the top. Side stakes, straps or lashing materials should be used on high loads where there is a danger of toppling. Maximum load limits should be posted on all equipment, including cranes and hoists. Overloading should not be permitted under any circumstances. Equipment should be safely unloaded from the top down.

Often the size and shape of a particular load requires the use of special attachments on existing equipment. The development of forklift truck attachments to handle special loads has increased the versatility of the forklift truck. Special attachments such as rams, detachable cranes, and fork extensions are available. In most cases, their use modifies the rated capacity and imposes weight limitations. Any modifications must be considered when such attachments are used to handle excessively large or bulky loads. Only attachments approved by the forklift manufacturer should be used. Overhead guards should be provided on all high-lift forklifts and other equipment where operators are in danger from falling objects. Overhead guards should be maintained free of any material that may obstruct an operator's vision.

Safe operating speeds should be set for each type of special-purpose handling vehicle. No vehicle should exceed 8 km/h (5 mph) inside warehouses or other buildings, regardless of its safe operating speed. On long hauls out of doors, speeds may be increased (within local speed limits) to the vehicle's safe maximum. Operators should use common sense to determine how fast they should safely travel with particular loads. The operator is responsible for the safe operation of the vehicle and transport of the load.

Loaded vehicles should not be put into motion until the load is properly stacked and secured. Drivers should be particularly careful when approaching doorways, aisle crossings, and other intersections; they should be observant at all times. Ramps, doorways and other physical features of warehouses and loading docks present a hazard to vehicle operators. Where possible, the conditions should be altered to reduce the hazard, or handling procedures should be revised.

When it is necessary to park vehicles inside warehouses, leave them only in an approved area, which has been authorized by the supervisor having jurisdiction. When a vehicle is parked, its brakes should be set, the ignition turned off, and the keys pulled. Unless the vehicle is equipped with an adequate passenger seat, and specific authorization is given, no person other than the driver should be allowed to ride on any materials-handling vehicle.

Fueling forklift trucks

Materials-handling vehicles should not be fueled indoors unless NFPA guidelines are followed in their entirety. (See References, NFPA 30.) Such vehicles should not be fueled in areas where flammable vapors may accumulate or where hazardous materials are stored. Fuel should be transferred only at designated safe locations, and the ignition turned off during fueling. At least two Class 20:B:C fire extinguishers should be kept ready during all fueling operations in case of a fuel spill or fire.

Precautions should be taken to prevent a fuel spill. If a small spill does occur, it should be neutralized, completely removed, or allowed to evaporate before a vehicle's engine is started.

Whenever large amounts of fuel are spilled, the fire department should be notified immediately to remove the hazard. While waiting for the fire department, materials-handling personnel should stand by the spill with suitable firefighting equipment. Where fuel-filling pipes are located close to the engine manifold or another source of ignition, the vehicle should be shut down and the engine allowed to cool before it is fueled. Much materials-handling equipment has been set on fire by spilling fuel on hot engines or exhausts. No vehicle should be fueled inside a building unless NFPA guidelines are followed in entirety. During servicing, the operator should turn off the engine.

Rules for operating materials-handling vehicles

The following section lists general standards of safe operation for materials-handling vehicles. All operators should be required to observe these standards.

- Before using any equipment, the driver should be properly trained in its use and should be listed as qualified to operate the vehicle.
- Before operating it, the driver should check the operability of the machine, using a checklist.
- Operators should drive at speeds that permit complete control at all times. Posted speed limits should be strictly observed. Drivers should be particularly careful when operating on slippery surfaces.
- Materials-handling vehicles should not be shifted from forward to reverse without coming to a complete stop.
- Riding clutches excessively abuses vehicles. Loads should be carried in the gear that allows the vehicle to move smoothly without excessive use of the clutch. Generally, the heavier the load to be moved, the lower the gear to be used.
- Drivers should make certain that their vehicle has sufficient clearance through doorways, at loading docks, in aisles, or where turns are made. Before passing through a doorway, operators should check for adequate clearance on all sides and at the top of a load. They should make certain that overhead doors are raised to maximum height before entering or leaving.
- When entering or leaving a building, or at intersections where the view is restricted, vehicle drivers should stop, sound the horn or bell, and proceed with caution.
- Drivers should be prohibited from carrying loose materials on any handling equipment.
- Operators should not be permitted to eat, drink or smoke while driving, nor engage in any unnecessary conversations.
- Drivers are responsible for safe delivery of loads. Unsafe loads should not be hauled. If a driver believes any load cannot be moved safely, the driver should immediately inform the supervisor, and then follow instructions.
- No person should be traveling on any vehicle unless a seat is provided for him or her (i.e., "No seat – no ride").
- Materials-handling vehicles should be parked in spaces approved by the appropriate authority. No vehicle should be parked in front of fire alarm boxes, fire extinguishers, safety equipment, or in aisles or other areas where it could become a safety hazard or interfere with emergency equipment. When parking equipment with gas tanks mounted higher than the carburetors, fuel should be turned off at the tanks, particularly when vehicles are parked inside warehouses. Preference should be given to the use of battery-operated forklift trucks inside buildings whenever practical.

Forklift trucks. Forklift truck operators should travel with forks close to, but not touching, the floor. With the forks close to the floor, the vehicle is less likely to hit and damage materials if it turns sharply. When traveling with a load, drivers should keep the forks close to the floor to prevent their field of vision from being obstructed by the material being transported. On an incline, the load should be kept on the uphill side.

Forklift trucks should be parked only in specified areas off the aisles. Forks must be lowered flat on the floor, the operating levers left in neutral, the ignition turned off, keys removed, and the hand brake set. This procedure should be followed at all times.

Drivers should always face or look in the direction in which they are moving. When moving a load too high to see over, operators should travel in reverse to give themselves a clear field of vision.

Gasoline-powered forklift trucks must never be left unattended with the motor running. This wastes fuel and creates excessive carbon monoxide concentrations if in confined areas.

Operators should not operate their trucks with wet or greasy hands. Serious accidents can result if a driver's hands slip on the steering wheel and control of the vehicle is lost.

Operators should not permit any person or persons to ride on the forks or other parts of their vehicles. Forklift trucks are not built to carry passengers. However, in cases where permission has been granted to use a forklift truck to lift an operator to the top of stacked materials, an approved platform and cage or safety pallet should first be securely attached to the lift. The person being raised should take a safe position where he or she is in no danger of falling or catching hands or clothing on the truck as the cage moves upward. A proper safety harness and lanyard should be used. Persons should not be transported from one place to another in a safety pallet.

Forklift operators should use general highway driving rules indoors as well as outdoors, keeping to the correct lane, stopping at blind intersections, and using horn and arm signals. Horns should be used for necessary warning only. Horns should not be blown to attract the attention of other operators.

At least three truck lengths should be maintained between moving vehicles to reduce the potential of collision or toppling a load if the front truck should stop suddenly. Operators should keep their legs within the vehicle when it is moving. Serious injuries have been caused when drivers' legs have been pinched between the truck and obstacles in narrow operating areas.

Forklift truck drivers should not carry empty pallets stacked higher than eye level. This can cause an unsteady load, and is a possible source of injury or property damage. Forklift trucks should be conspicuously marked with their rated lifting capacity.

Tow tractors. Tow tractors are used for a variety of materials-handling operations, particularly where extremely heavy loads are to be moved considerable distances. The following precautions should be observed in tow tractor operations:

- Passengers should not be allowed to ride on tractors unless adequate seats have been installed.
- No more than the recommended number of trailers (usually four), loaded or empty, should be pulled by any tractor. In order to avoid jackknifing, trailer trains should be arranged with the heaviest trailer load next to the towing vehicle, the next heaviest second in line, and so on.
- Tow tractor operators should make certain that couplings are secure before moving a trailer or train.

This can be achieved by a visual inspection and walkaround.
- Tow tractors with trailers in tow never should exceed 8 km/h (5 mph).

Cranes and hoists

Lifting devices include a large group of cranes, derricks, hoists, elevators, and jacks. Slings, ropes, chains, and cables are used with lifting machines or as separate lifting aids. Lifting device hazards include swinging or falling loads, exposed moving parts, structural failures, and snapping cables and ropes.

All cranes and hoists should be installed, maintained, and inspected in accordance with applicable governmental requirements. Many lifting device failures can be prevented if proper inspection, maintenance and operating procedures are followed. Operators on each shift should inspect lifting devices and lifting aids for operational condition and defects. When such equipment is not assigned to one operator, designated individuals should be responsible for making a daily inspection. Conditions that make equipment unsafe for operation should be reported immediately, equipment should be locked out of service, and repairs should be arranged.

Before any crane or derrick is moved, the route of travel should be checked to be certain that adequate clearance exists along the entire route. Operators should be particularly alert for overhead power lines. During actual movement, the hook should be secured against swinging and the boom lowered. A red warning flag or light should be placed at the exposed end of the boom. Heavy materials should not be carried at the end of the boom while equipment is being moved from one location to another.

- All cranes and derricks should be kept away from walls, overhead trestles, columns, and other structures.
- Safe load limits and boom angle indicators should be installed in plain sight on all cranes and derricks, for both maximum and minimum positions. These position load limits should not be exceeded.

- Cranes and derricks should be equipped with braking devices capable of stopping a weight of at least one and one half times the rated maximum load. "Deadman" type braking devices are preferred.
- Ladders, steps, handholds and similar devices should be provided for safe access to cabs. Platforms with railings and toeboards should be installed for operators' protection.
- All exposed machinery and dangerous moving parts should be guarded against contact by personnel.
- Under no circumstances should personnel be permitted to ride loads being moved.
- At least two full turns of cable should remain on hoist drums at all times.
- Booms, buckets or other loads should not be moved over personnel below.
- Booms should be placed directly over loads when lifting. If needed, snatch blocks should be used to move loads into position under the boom.
- When outriggers are to be used, they should be extended fully. For rubber-tired mobile cranes, the outriggers must be extended fully, elevating the tires completely off grade to reduce possible turnover. The crane should be leveled as for other operations.
- The swing radius of the crane should be guarded to prevent employees from being injured by the crane counterweight or by pinch-points.
- Tag lines should be used on free swinging loads to help guide them and to prevent them from striking nearby objects.
- Booms not in use should be lowered to the ground, or onto resting platforms, to remove tension from cables and drums.
- Except as specified by maintenance requirements, no maintenance should be performed on cranes or derricks while they are operating.
- No part of a lifting device (particularly the boom) or its load should be permitted to come within 3 m (10 ft) of live electrical power lines rated 50,000 volts (50 kv) or less. Greater clearance is required for power lines over 50 kv.

NOTE: WHERE THE LOCATION OF THE OPERATION MAKES THIS IMPRACTICAL TO KEEP CLEAR OF ELECTRIC LINES, ELECTRIC CURRENT SHOULD BE SHUT OFF.

A PERMANENT SIGN SHOULD BE PLACED IN THE CAB OF THE EQUIPMENT IN FULL VIEW OF THE OPERATOR, READING: "DANGER: DO NOT OPERATE THIS EQUIPMENT WITHIN 3 M (10 FT) OF ELECTRICAL POWER LINES OF 50 KV OR LESS." FOLLOW YOUR LOCAL, STATE OR FEDERAL AUTHORITY; OR YOUR COUNTRY'S NATIONAL STANDARDS BODY.

- If the boom of a crane should contact a hot wire, the piece of equipment will become energized and the circuit breaker may not operate. When this happens, operators should not step to the ground as their bodies may complete a circuit. If the crane is functioning, move the crane away from the hot wire. If it is absolutely necessary to leave the equipment, operators should hop clear of their vehicle, keeping feet together.
- Cranes should be equipped with positive stops to prevent running off the ends of jibs, rails or bridges.
- Only authorized, trained personnel should be permitted to operate cranes and derricks.
- No individual should be allowed to perform repairs on crane booms unless authorized by manufacturer's service representatives or certified by the company as a crane boom repair specialist.

In general, all lifting or overhead carrying devices should be fitted with adequate equipment to protect operators from falling objects, swing loads, and cable failures. An efficient, audible warning device should be provided when hoisting operations or moving equipment might endanger people working in the area.

Appropriate head protection should be worn by crane crew personnel when such protection is needed to protect operators from falling objects or impact hazards. Safety-toe shoes should be worn by all operators involved in crane, hoist, or derrick operations.

Monorail hoists should not be used to move an object by a side pull, unless properly designed for that purpose. Monorail hoists operated in swivels should be equipped with one or more safety catches that will support the load should a suspension pin fail. Both the track and its supports should be inspected frequently for signs of weakening and wear. Rail stops should be provided at the ends of a monorail track. Such rail stops should extend at least as high as the radius of the wheels.

At switches, turntables, and transfer tables, automatic bumpers should drop into position to prevent the trolley from running off the open ends of the fixed and movable track if they are not properly lined up with each other. Conversely, the track should be interlocked with the bumpers so that the track cannot move until the rail stops are in position. Maximum load capacity should be plainly marked on each side of a monorail, and should be clearly legible from the ground or floor.

REFERENCES

American National Standards Institute:
Identification of Piping Systems, Scheme for the, ANSI A13.1-1996.

Jacks, ANSI/ASME B30.1-1998.
National Fire Protection Association:
Aircraft Hangars, NFPA 409, 1997.
Aircraft Maintenance, NFPA 410, 1997.
Flammable and Combustible Liquids Code, NFPA 30, 1997.
National Electrical Code, NFPA 70, 1997.
Spray Application Using Flammable and Combustible Materials, NFPA 33, 1997.
National Safety Council:
Accident Prevention Manual for Business and Industry.
Industrial Data Sheets:
Lead-Acid Storage Batteries, 635.
Mounting Heavy-Duty Tires and Rims, 411.
Portable Ladders, 665.
Lifting Calculator.

Underwriters Laboratories Inc.

U.S. Department of Health and Human Services, National Institute for Occupational Safety and Health (NIOSH). *Work Practices Guide for Manual Lifting, NIOSH Technical Report 81-122.*

CARGO OPERATIONS

CHAPTER 19

Truck dock operations
Dock areas
Checking cargo
Loading and unloading trucks
Handling cargo
Forklift operations
Cargo handling systems
Live-roller conveyors
Tow conveyors
Storage racks
Loading and unloading procedures for aircraft
Cargo carts
Cargo pallets
Transporting pallets
Loading cargo planes
Traveling lifts
Cargo aircraft without mechanical loading systems
Wide-body and supersonic aircraft
References

Safety and health laws directly and indirectly control international air cargo operations. Because the air freight terminal is a 24-hour-a-day complex with hundreds of employees, mechanized processing and loading systems, and services to suit every shipper and industry, incident prevention is essential in running a smooth, profitable operation. (See Figure 19-1.) If passengers and cargo are on the same plane, special safety procedures must be observed. The key to greater safety is training: A well-trained work force is the most valuable asset.

It is the purpose of this chapter to outline the accepted practices, facilities, and equipment that contribute to making freight operations safer and more efficient. Where necessary, reference is made to laws and standards that regulate operations.

TRUCK DOCK OPERATIONS

This section covers procedures that should be established in dock areas to process cargo safely and efficiently, both manually and mechanically.

Dock areas

The freight yard and dock area should be clean and well-illuminated so that all vehicles can approach them safely. Employees should be instructed to stay away from areas in front of docks to avoid injuries. Trucks should be positioned flush against the dock to prevent personnel from falling through gaps. Truck wheels require constant chocking. Open dock edges, which are not used for truck mating, should be equipped with a suitable anti-rolloff device. Fixed ladders should be provided between dock positions to enable truckers to gain easy access to docks.

Physical barriers should be positioned to prevent the public from entering work areas. Shippers and truckers should be restricted to the office area and should not be allowed to wander around the air freight terminal. Appropriate warning signs should be posted and the restrictions enforced.

Portable dock plates used as bridges between trucks and docks should be of a design and strength appropriate to that function. Standard plates have a device to prevent any movement. Whenever possible, plates should be moved by forklift trucks rather than manually, to prevent back and hand injuries to the operator.

A preventive maintenance program should be in place to ensure all door mechanisms are serviceable. Periodic inspections should ensure that the doors are not likely to fall, causing injury or damage. Additionally,

Figure 19-1. Loading areas should be clear of all unauthorized vehicles. Equipment should be parked in designated areas a sufficient distance from service ramp to prevent damage to aircraft or vehicles.

Photo courtesy of China Airlines

truckers whose vehicles have long, overhanging loads should be instructed to stay out of the dock wells until the door is opened.

Fire extinguishers should be provided to protect against truck engine fires, as required by the National Fire Protection Association (NFPA 10; see References). Extinguishers must be inspected periodically according to applicable laws or as required by your local, state or federal authority; or by your country's national standards.

Dock floors should be "broom clean" and free of any foreign material. Operators should be particularly alert for any items on floors that may damage freight or cause injuries. Cracks in floors should be filled with patching material to prevent tripping or catching crates that are slid over the surface.

Electrical outlets and equipment provided should meet applicable building codes; grounding should meet all regulations.

Checking cargo

For safety, security and service reasons, cargo should be carefully checked when received or delivered. Packaging should meet standard specification as required by your local, state or federal authority; or by your country's consensus standards. Most of these standards are based on the International Air Transportation Association (IATA) Dangerous Goods Regulations and/or ICAO Technical Instructions. (In the United States, Department of Transportation (DOT) regulations apply; see 49 CFR 173.) Special care should be taken if dangerous goods are involved. Dangerous goods cargo initially not accepted for shipping can be repackaged in accordance with applicable standards and then processed for shipping. Salvaged packed shipments should be entered onto the shipper declaration (air waybill).

High-density shipments should be provided with skids or pallets to spread the floor load and prevent damage to an aircraft floor. Operators should be alert for cargo with sharp projections or edges, or with unnecessarily long bolts or nails.

When accepting livestock – particularly wild animals – operators should make sure cages are secure and will withstand rough handling. Employees should be cautioned to avoid allowing their hands to enter animal cages. Shipping animals capable of inflicting injury or damage during transit should be carried out in accord with the International Air Transport Association (IATA) standards. (See References.)

Some commodities require special packaging to protect them from damage during ordinary handling (e.g., dry ice used to preserve a perishable commodity). Check governmental/local standards and company regulations or procedures to determine if packaging is acceptable. A commonly used reference in the United States is 49 CFR, 173.

Damaged inbound shipments should be repackaged before delivery to customers or truckers to prevent injury or loss by spillage or theft. A damage report should be prepared as required. Also, international regulations for the reporting of dangerous goods incidents may apply. In the United States, if dangerous goods are involved, notification at the earliest practical moment is required to the nearest FAA Civil Aviation Security Office, and a Hazardous Materials Incident Report needs to be submitted. (Such forms may be obtained from state DOT offices or the Office of Policy Development and Information Systems Division; see References.) Notification may also be required by the Environmental Protection Agency immediately after release of a reportable quantity of any hazardous substance or pollutant.

Loading and unloading trucks

Cargo on trucks should be checked to see that it is stowed properly to prevent spilling or shifting in transport. Heavy packages and wooden crates should be placed on the bottom; light packages and cardboard containers, on the top. The height of storage racks should not be exceeded.

When cargo is hand-loaded from the truck to the dock, pallets or hand carts should be positioned as near as possible to the tailgates of the trucks. Only serviceable pallets, free of broken boards or projecting nails, should be used.

When forklifts enter a truck to unload large pieces of palletized cargo, drivers should make sure that truck brakes are set and that truck wheels are chocked on both sides to prevent movement of the truck away from the dock. Signs should inform truckers of these requirements. Forklifts should enter only those trucks with bed capacities well above the weight of the forklift. Air freight terminal supervisors should check for compliance with this procedure. Forklifts should not be driven into a truck trailer without first inspecting the floor to ensure structural integrity of the floor.

Hand carts should be provided with brakes to prevent their rolling off a dock. Hand trucks, pallet jacks and similar equipment without brakes should be kept away from the dock edge to prevent them from rolling off. Forklift trucks and other self-propelled equipment should have brakes set when a driver dismounts.

Handling cargo

A flow pattern should be planned to minimize the number of times cargo is handled. Whenever possible, cargo should be loaded on the pallet or cart on which it will be stored. Piles should be stacked evenly and "cross-tied" to bind the stack together, particularly in terminals where loads are stacked two or more tiers high. If possible, stacking pallets with loads not bound by wire or steel bands should be avoided.

Odd-sized or top-heavy shipments should be laid flat to prevent their falling over. Those that, by their nature, cannot be laid down should be secured to a base or pallet with the correct side up, as indicated by the imprinted arrows on the pallet. Small pieces should not be left on a floor where they will be a tripping hazard while building up a stack.

When manually moving cargo, operators should use extreme care to avoid injuries. Every effort should be made to use mechanical lifting methods when moving cargo, but when an unusual handling operation is unavoidable, the risks should be identified and an appropriate level of care exercised. Training in proper lifting methods is vital. For example, operators should be instructed to avoid twisting their backs; instead,

they should move their feet in the direction of the lift. They should avoid sudden or jerky movements. When lifting heavy or awkward pieces, operators should keep their backs aligned with the natural posture curves and bend their knees. The load should be kept as close to the body as possible. Before handling any load, the operator should determine if help is needed. Two or more operators should be used, as required. When more than one person is lifting a load, all movements should be coordinated – they should lift or lower together to prevent one person from holding all of a load. Cargo should never be thrown from one operator to another.

Operators should wear gloves when handling wooden crates, metal parts, castings or sheet metal parts. They should not pick up shipments by their bands or ropes. Safety shoes should be required for those dock operators who are exposed to toe and foot injuries. It should be a mandatory requirement to use protective clothing or equipment when working with known hazards.

Forklift operations

Forklift operations can be hazardous, and they require careful supervision. Operators must be appropriately trained and at regularly scheduled intervals receive refresher training that is relevant to the task and environment in which they are employed. The forklift operator should receive a permit or other documentary authority for forklift operations with records maintained by the department to ensure validity. A set of rules for operating forklifts should be posted in the work area, and their use should be closely supervised. Operators in the area should stand clear when pallets are being moved by forklift trucks to avoid being crushed or having their feet caught. Drivers also should be careful when inserting tines in a pallet opening to avoid pushing a pallet. When loading pallets in racks or carts, operators should observe clearances on all sides to prevent damage to cargo or equipment.

Operators can avoid serious injuries from tipovers by following prescribed safety guidelines and by wearing a seat belt. There are many causes of tipover.

Operators must remember that lift trucks have small, narrow wheel base (much smaller than a car) – unlike a car, the weight/load is being carried outside of the lift truck. The heavy counterweight is intended to keep a balance between the load and the back of the lift truck. Speed often is the cause of major incidents as well as tipover.

Forklift tines should be spread as wide as possible to increase load stability. A load should be tilted back slightly before it is moved. Forklifts should move slowly in close quarters, and avoid fast or sharp turns. Operators also should be on the alert for other employees who may not see or hear them approaching. When the load obstructs the operator's vision, the forklift should be driven in reverse.

When placing pallet loads in racks or one on top of another, operators should make sure the pallets are level or are resting on the cross members of a rack. When stacking one loaded pallet on another, operators should be guided by the weights involved, the rigidity of the packages, and the nature of the cargo. Internal combustion engines should not be operated in a confined space unless there is adequate ventilation to remove exhaust gases. Electrically powered forklifts might be considered.

CARGO HANDLING SYSTEMS

Systems in air freight terminals differ little from those in surface carriers' terminals. The principal difference is that in air freight systems cargo is often sorted and stored in small lots for destinations that have little or inadequate mechanical handling equipment.

Live-roller conveyors

All electrical wiring on roller conveyors should be protected in thinwall or flexible metal sheathing and meet local building codes or as required by your local, state or federal authority; or your country's national standards body as well as the National Electrical Code (NFPA 70; see References). Belts, pulleys, sheaves and drive systems should be guarded to prevent nip injuries.

When cargo is hand-sorted or transferred to carts or pallets, the conveyor should be at a convenient working height to reduce lifting. Conveyors should not be overloaded. Maximum weight limits for single pieces should be observed as specified by conveyor manufacturers. Signs that show maximum weights should be posted at the starting points on systems.

Signs stating "Walking on rollers is prohibited" should be prominently posted and the rule enforced.

Removable stops at the ends of all conveyors should be provided to prevent cargo from falling off. Employees working at these ends should be alert to prevent cargo from falling on them. Conveyors should be protected with curbing if mobile equipment is used alongside them.

Live rollers are driven by belts located below them. These belts are a hazard to anyone below the conveyor. Before maintenance is performed, or anyone is allowed to crawl under one to retrieve fallen items, the conveyor should be shut down and its power locked in the "off" position. (OSHA regulations require a formal lockout/tagout program for operators and maintenance.)

Tow conveyors

There are three types of tow conveyors: overhead chain, in-the-floor and magnetic track. The latter uses a battery-powered tractor and follows a track that can be moved at any time.

The first two are continuous chain conveyors and are mechanically driven. Both of these systems should have emergency shutoff controls strategically located around the terminal. The shutoffs should be prominently placarded and all employees instructed in their use. A warning signal should be sounded before restarting.

Overhead conveyors. For overhead conveyors, dragline paths should be painted on the floor to indicate the path of handcarts so that no obstruction will be in their way. Only standard handcarts designed for draglines should be used. No ordinary handcarts or dollies should be hooked in. To prevent being caught and dragged, operators should stand clear of chains on overhead systems when the chains are not attached to carts. To prevent damage to the overhead conveyor, forklift trucks should not be allowed inside an area of a dragline.

In-the-floor conveyors. In-the-floor conveyor systems are more common than the overhead type. The freight terminal area should be kept clear of any material that may jam into a drag slot embedded in the floor. Again, only standard carts should be used and only a standard drop rod should be used to connect carts to a chain.

Magnetic track conveyors. Magnetic track systems are most flexible; they can be moved by merely picking up the track or turning the track in another direction. These tracks should be moved only in accord with predetermined layouts so a new routing will not create a hazard. Only shift supervisors should have the authority to relocate the flow pattern. Consideration should also be given to the proximity of the track system to dock edges, fixed equipment, storage racks and other hazards.

Standards for conveyors. All handcarts should be provided with brakes to prevent their rolling when removed from draglines. There should also be a maintenance program to ensure safe, reliable operation of all systems. This should include lubrication, battery charging, handcart repair and periodic checks of drag chains. A positive lockout is also needed for repair work on these systems.

Storage racks

Numerous kinds of cargo-storage racks are available. Lightweight racks with patented locking devices are frequently used because of their flexibility. These, on occasion, have not stood up under the usage common in air freight terminals.

Wedge-locked devices should not be used. They can be unlocked simply by lifting the main supports and can collapse if locks are incidentally lifted by forklift trucks. Bolts or spring-loaded pins cost more but are far more reliable. Locking devices should be checked to ensure they are intact.

There should also be a sufficient number of cross braces to prevent racks from spreading. Cross braces should be placed to help support the pallets and prevent tipping. Rack width should be consistent with the pallets used. When pallets are stored in a rack, their bottom end boards should be outside the main frames to prevent the pallets from slipping. Rack weight limitations should be posted or stenciled on all racks. Rack members should not protrude beyond the front vertical supports.

Aisle width between racks should follow the recommendations of forklift truck manufacturers and applicable safety regulations. Aisles that are too narrow will result in incidents, damaged racks and cargo, and decreased efficiency. Convex mirrors should also be installed at blind intersections. Forklift truck drivers should be trained to operate in close quarters. Supervisors should periodically monitor forklift operators and, if necessary, correct their driving habits and their methods of pallet stowage.

A wide variety of forklift trucks is available. It is important that the right type for the terminal be purchased, since proper equipment is a key to a safe, efficient operation. Forklift trucks should have sufficient capacity to handle maximum pallet loads. The maximum lifting capacity of the truck (in relation to the load centers) should be displayed in view of the operator. Also, if the lift truck is operated in an area where falling pieces of freight are a hazard, overhead protection should be provided on the forklift.

Because of the hazard of falling cargo or collapsing racks, supervisors should constantly monitor pallet stacking, transporting, and stowing. Supervisors must also monitor vehicle operation in the terminal to make sure that correct driving practices are being followed. Speeding or other dangerous driving within a terminal area must not be tolerated. Indoor speeds should be walking speed.

LOADING AND UNLOADING PROCEDURES FOR AIRCRAFT

Cargo is transported in two kinds of flights: flights that carry cargo only, or flights that carry cargo and passengers. The movement of cargo to and from gate areas has become one of the most important phases in cargo handling. A breakdown in loading cargo will affect the entire operation – passenger service as well as cargo service.

Assignment of aircraft space for cargo loads is a function of a dispatch group. Safety responsibilities of this group should include checking cargo for correct content, packaging, weight, size, condition (will not cause injury or damage to aircraft) and acceptability for air transport.

It is important that only the freight recorded on the manifest be processed for loading. Sloppy loading can result in overgrossing an aircraft, incorrect balance, or possible injury to employees or passengers from unacceptable freight. Checkers must work closely with load-makeup personnel to make sure that the correct freight is loaded in the correct carts and that freight requiring special handling receives it. Carts should be identified by flight number for correct loading at gate areas.

Cargo carts

When loading cargo carts, the loading crew should stow freight evenly, with heavy pieces on the bottom and in the center to ensure stability. (To minimize lifting, the loaders might consider positioning heavy pieces next to the gate that will be used to unload.) They should make sure that the weight capacity of these carts is not exceeded. In addition, oversize pieces should not extend over the sides because they could endanger personnel or damage equipment. All doors, gates and curtains on carts should be secured to prevent freight from falling out.

When hitching carts and tractors, operators should make sure hitches are serviceable and securely latched. Operators should straddle the tow bars when hitching, to prevent incidental drops. Connecting loaded carts should require the efforts of two employees.

Tractors should never be used to push carts when someone is in front of the cart. Carts must be under control at all times.

When towing cargo carts, drivers should slow down around corners to prevent cargo from shifting and overturning the carts. Most cargo cart trains tend to "drift in" (shorten the turning radius) when turning corners. For example, in a three-cart train, the last cart may hit the door frame even though the first clears it. Therefore, drivers should avoid turning prior to, or immediately after, passing an obstacle.

When turning a cart train, drivers should stay clear of trailing edges of wings or any projections. They should also be alert for fueling hoses and hydrant valves (Figure 19-2).

Figure 19-2. Trains of loaded cargo carts should move at a maximum speed of 16 km/h (10 mph). Because most cargo trains "drift in" during turns, drivers should avoid turning immediately before or after an obstacle. They also should stay clear of the trailing edges of airplane wings or any other projections.

Photo courtesy of Federal Express

Management should contact the airport authority to determine the maximum number of carts to be towed in a train. Factors to consider in the decision are equipment specifications (e.g., tongue bolt strength, trailing characteristics, etc.), airport authority regulation and prevailing operational conditions.

Empty cargo carts on ramps should be available to unload inbound cargo and prevent double handling. Cargo carts should be parked within designated parking lanes and their brakes set.

Ramp personnel should be made responsible for loading cargo correctly in the carts, and securing all gates and doors on the carts. Drivers should be assisted in hitching carts to tractor trains. In some places, one group of drivers transfers cargo between terminals, while another group takes over operations on ramps. The ramp services group should be responsible for moving the carts to a staging area and for preventing exposure to jet blasts of nearby aircraft.

Because of the possibility that the ramp service crews may have loaded inbound cargo incorrectly into the carts, freight terminal crews should exercise care when opening cart gates or curtains. They also should check the stability of the stacking and make sure pieces are not leaning against cart gates. Cart gates should be lowered carefully. Operators should not turn their backs on a full cart, because serious injuries have resulted from freight tumbling out of carts. Operators should unload from the top down, without dislodging any "key" boxes.

Because of the danger of sprained ankles from feet slipping between the deck boards, operators must be prohibited from walking on pallets when processing freight. When moving freight from carts to pallets, operators should stand at an angle to the load. They should pivot by moving their feet to avoid twisting. Pallets should be stacked carefully to prevent spilling.

Broken cartons should be repackaged at once. Damaged freight never should be placed in racks for transfer or delivery. Notify the freight service or sales group about shippers that consistently use poor packaging.

All cart gates should be closed when the carts are emptied, and the carts should be moved out to a parking area. Brakes should be set when the carts are parked. When towing in a terminal, drivers should follow the painted lanes and avoid short radius turns.

Cargo pallets

Pallets are used in most cargo aircraft systems. These vary in size, according to carrier, but usually conform to industry-standard sizes. The most common size is 2.2 m x 3.2 m (88 in X 125 in). Pallets are made of aluminum or hardboard with a metal edge. Fixtures in the metal edge provide attachment points for tying down nets.

These pallets can be built up:

• By the shipper. The pallets are transported to the freight terminal by flatbed trucks or ramp equipment
• By another carrier
• At the dock, from trucks directly to pallets
• From storage to pallets

Whichever method is used, freight on pallets should be secured to withstand movement. These restraints are specified by government regulations (DOT in the United States). Two means can be used to meet these requirements – nets or containers.

Restraint with nets is accomplished by lowering the net over the pile of freight and adjusting the length of the straps by buckles or hooks. Nets should be sufficiently tight to prevent freight from shifting (Figure 19-3). Specific weight limits apply to palletizing and netting cargo. Pallet loads that shift can damage the interior of the cargoliner during loading and unloading, as well as during flight. Vertical and horizontal forces that may occur during takeoff, landing or in flight as a result of turbulence must be considered when securing loads.

Containers may be closed on all sides or open on one side. The latter type is contoured to fit the fuselage and is referred to by trade names such as "huts" or "igloos." When loaded on an aircraft, the open sides of containers must be facing the same direc-

Figure 19-3. Cargo that is on a pallet may shift unless it is held in place by the right kind of net.

Photo courtesy of United Air Lines

tion. Any unused space in the containers may be filled with inert dunnage to prevent the cargo from shifting during flight.

All containers should be secured to pallets and should meet the necessary restraint requirements. It is important that restraints be secured to a pallet at all attachment points to prevent loads from shifting during flight.

Some container bases have little rigid strength. Before handling them, determine whether they can be handled with a forklift truck. If container bases cannot be moved with a forklift truck, then they may be moved only on conveyor systems or on transporters equipped with conveyors. Lifting them with forklift trucks or allowing them to roll off conveyors will result in broken and distorted containers.

Loads should be distributed over an entire pallet; concentrating a heavy load in the center or in one small area will cause difficulty in loading the aircraft, and

may damage the conveying system and prevent some of the locks from engaging the pallet. Improperly locked pallets have broken free during flight, inflicting major damage on aircraft cabins.

The maximum allowable weight for pallets specified by the manufacturer or company procedure should not be exceeded. Overloading could result in an incorrectly balanced aircraft. Therefore, for safety and efficiency, the entire pallet container and load should be weighed before dispatching.

Transporting pallets

Pallets usually are transported between the air freight terminal and the aircraft on a variety of vehicles. All have roller conveyor beds, some with power. It is most important, therefore, that end locks and side guides are installed on conveyors so that the load cannot roll off incidentally. Locks left in the down position when transporting cargo are a major source of incidents and damage. Special emphasis should be used in this area. The locks should be checked daily. Employees should be instructed that any mechanical cargo transporters with unserviceable parts must be tagged and immediately taken out of service. When overhead cranes are used to move pallet loads, the area should be restricted to authorized personnel only.

Transporter dollies are drawn by tractors or powered tow bars. Transfer by powered tow bars is a very simple operation, but has resulted in many injuries. Operators should be given a complete course of instruction by manufacturer's representatives or by company instructors. The most serious hazards are:

- Operators backing themselves against a wall
- Machines running over an employee's feet
- Excess speed
- Personnel falling under a power unit or dolly
- Breakaway dollies from poor or improper hook-up
- Controls being activated in the wrong direction
- Losing cargo from dollies
- Operators backing dollies into employees
- Failure to stop in time
- Failure to allow sufficient clearance, and striking

equipment or fixed installations
- Not pre-tripping the tractor for possible mechanical defects
- Not reporting mechanical defects

Supervisors should observe periodically the operation of powered tow bars and take corrective action as needed. Operators must obey safe practices.

Batteries on tow bars should be charged in accord with governmental, local, and industrial standards. Only trained personnel should be allowed to do this work. Battery-charging areas should be isolated, if possible, to reduce potential damage to other equipment and should be inspected regularly. Smoking should be prohibited in the area.

Transporter trucks are basically flatbed trucks with conveying systems and mating devices to connect transporters to storage racks or aircraft loaders. Elevating trucks should have guardrails and fully covered decks. Trucks also require end stops, whether the rollers are powered or not. Trucks with any unserviceable controls should be immediately tagged "out of service."

Vehicles should be driven at low speed regardless of distance involved because of the height of the load and the method of restraint. Here, too, end stops left in the down position have contributed to many incidents and serious cargo damage. Operators should be particularly alert for other personnel working in the area because a vehicle with a load is completely blind on one side.

Loading cargo planes
Never approach an aircraft with load equipment until the following has been completed.

- The aircraft has come to a complete stop and is in a parked position
- Engines are shut down
- Wheels are chocked
- Red beacon light is turned off
- Tail stand is in place (on aircraft equipped with such)
- The marshaller gives the signal to approach

Two basic loading systems are used for cargo aircraft: portable loaders that are driven to the aircraft, and loading docks or bridges against which the aircraft parks.

A number of portable loading units are available; these recommendations apply to all loaders.

- Vehicles should have brakes capable of stopping or holding them under any operating conditions.
- Vehicles should operate at slow speed.
- A person acting as a guide is required when loaders are being positioned against airplanes.
- Operators should make sure that no part of a loader touches an aircraft, either when lowered or when raised to the door level.
- The following devices should be standard equipment on loaders: stabilizers, parking brakes, warning devices, automatic pallet stops, flow-control orifices (to prevent lifts from falling suddenly if power fails), guardrails for employee protection, weatherproof controls and placards on all controls.
- When maintenance is being performed under an elevated loader bed, mechanical guards should be used to prevent incidental collapse of the bed.

Fixed bridges have certain unique hazards of their own. These procedures should be followed to ensure safe operation of fixed bridges:

- Docking aids and taxi guidelines must be present so that aircraft can be positioned precisely.
- Bridges should be retracted all the way, which may also be necessary to prevent aircraft damage.
- Bridges must be extended carefully and accurately to provide correct mating with aircraft and to prevent damage (Figure 19-4).

Loading areas should be kept clear of all unauthorized equipment. No vehicles should be allowed outside equipment parking lanes. Employees meeting the aircraft must wear approved hearing protection.

When a portable loading system is used, equipment should be parked in a designated area that is a sufficient distance away from the service ramp to

Figure 19-4. Bridges must be extended carefully and accurately to provide correct mating with aircraft and to prevent damage.

Photo courtesy of Northwest Airlines

prevent the loader from being struck by improperly positioned aircraft.

Only after the aircraft has been parked and chocked should the loader be allowed to approach. A basic rule: Prohibit all personnel and equipment from entering an area while aircraft are arriving and departing.

Service personnel should enter the aircraft and open the main cargo door before a loader is positioned. The cargo door should be opened all the way and secured to prevent damage from wind. When a loader is operating, personnel should stand clear of the bed. In the event of a malfunction, the maintenance department should be notified immediately. Freight service personnel should not attempt any repairs.

Most loaders have numerous rollers or ball bearings on the deck. Operators must use care, therefore, to prevent falling or pinching their feet and whenever possible, should stay on the walkway provided. Walks should have slip-resistant surfaces for better footing. Guardrails should be installed.

When pallets are pushed into an aircraft, enough personnel should be provided to keep the load under control and to prevent operators from overexerting themselves. They should position themselves cor-

rectly to avoid jamming their hands or feet on door frames. When moving pallets into a stowing position, operators should monitor the shape of a load to make sure it does not damage the aircraft liner.

Pallets should be secured to the aircraft floor locks to prevent their shifting in flight. A pallet position checklist should be used by the lead person to ensure all locks are engaged as the aircraft is loaded and operators should keep their fingers clear of locks. If locks are inoperative, a mechanic should be summoned immediately. If some locks are not immediately repairable, certain load restrictions may apply. After the last pallet has been put in place, all locks in the doorway should be secured before the door is closed. Upon completion of loading, space permitting, the lead person should make a final walk through check of the cabin.

Unloading is, essentially, the reverse of loading. These general rules apply to both:

- Install a tail post as required during either operation to prevent the aircraft from "sitting on its tail."
- Report damage to interior liners in upper or lower compartments. The liners should be repaired to comply with governmental compartment classifications.
- When operators are pulling pallets toward the door of a plane, they should avoid catching their feet between the pallet and the floor hardware. Supervisors should make sure that employees wear safety shoes for this work, to avoid toe and foot injuries. Attachable pull straps should be available to assist.
- Operators should be alert for pallet loads that have become deformed in flight. Handling such pallets specially will help prevent liner damage.

Traveling lifts

Fixed docks or bridge systems use "traveling lifts" to move loaded pallets from storage to a bridge for loading in an aircraft. The lift operates on tracks similar to a railroad. To ensure safe operation, lifts should be designed so the controls for operating them and those for operating roller conveyors are not too far apart to be operated by one person at the same time.

The presence of two persons at the controls at any time should be forbidden.

The area in which a lift moves should be outlined with bold, painted lines, and a warning posted stating "Keep out of lift areas." Flashing lights and horns should operate whenever lifts are moving. Lift operators should be responsible for seeing that tracks are clear before entering terminals. Warehouse or terminal doorways must be wide enough to provide safe entry, if the lifts operate within.

Operators should inspect their lifts daily to make sure all controls, pallet stops, brakes and other safety devices are operating. Pallet dollies or transporters should have brakes set when positioned at the traveling lift for off-loading. Braking prevents their being pushed, which would allow a pallet to fall. Operators should check to see that pallet stops are up before loading pallets on dollies.

Cargo aircraft without mechanical loading systems

While aircraft with mechanized loading systems have assumed the spotlight in freight services, many aircraft in use do not have such systems. Many carriers, scheduled as well as nonscheduled, depend on them in their freight operation (Figure 19-5). Such aircraft are generally loaded by forklift truck, rather than by mechanized systems.

Wooden pallets used for loading aircraft must be serviceable. They should have a sufficient number of bottom boards to hold a load even when off center. Three-sided pallets provide better stacking and keep freight from falling. They also provide a means of stacking one on another without danger of crushing a load. Unserviceable pallets should be removed from service, and repaired or junked. This is particularly true of pallets with loose bottom boards. Fatal injuries have resulted from defective pallets that tilted and fell because boards were either missing or had pulled loose.

A pallet system depends on forklift trucks. The truck capacity required is determined by the maximum loads handled. All operators require special training before they qualify as aircraft loaders. Part of the training should convey an understanding of how the capacity of a truck is related to the distance of the center of the load from the front axle. An overloaded truck can drop its load or overturn. The maximum permissible load should be displayed in view of the forklift operator.

Drivers must make a full stop at least 3 m (10 ft) from an aircraft when loading. They should lift the load to just above doorsill height and move the pallet into a doorway slowly, making sure the loading crew is clear. They should also place pallets on the floor and then back straight out and not turn until the vehicle is past the wing tip or stabilizer. If there is insufficient room to leave a pallet in an aircraft, the leading board should at least be rested on the doorsill to stabilize the load while the forklift truck supports the load.

References

International Air Transport Association (In the United States: Cargo Network Services Corp.). *Dangerous Goods Regulations.*
National Fire Protection Association:
National Electrical Code, NFPA 70, 1999.
Portable Fire Extinguishers, NFPA 10, 1998.
National Safety Council:
Accident Prevention Manual for Business and Industry: Engineering and Technology, 11th ed., Chapter 14, "Materials Handling and Storage." 1997
U.S. Department of Transportation. Code of Federal Regulations, Title 49, Part 173.

Figure 19-5. While aircraft with mechanized loading systems have assumed the spotlight in freight services, many aircraft in use do not have such systems such as the one above.

Photo courtesy of Northwest Airlines

COLLECTION, MANAGEMENT, AND ANALYSIS OF SAFETY DATA

CHAPTER 20

Collection
Management
Content of the DTS
DTS infrastructure
Data input and extraction (retrieval)
Analysis
Descriptive methods
Rate calculations and graphical methods
Statistics
Inferential methods
Conclusion
References

This chapter addresses the collection, management, and analysis of safety data. It is important to recognize that this data and the associated information management system and analytical methods may pertain to safety events, but may also be used for any type of event or occurrence. Examples of events or occurrences are:

- occupational safety and health injuries and illnesses
- aircraft ground damage
- property (equipment and facilities) damage

Note that data analysis may be used for events and occurrences that "anticipate" a more severe event or occurrence. For instance, first aid, medical office visits, or near misses may be considered to anticipate injuries and illnesses.

Therefore, the purpose of this chapter is to provide information regarding the management and analysis of data collected in any type of situation. However, the topic will be approached from the perspective of safety.

Data is collected, managed, and analyzed for several reasons. The most obvious of these is to demonstrate compliance with regulations (for example, the U.S. Occupational Safety and Health Administration – OSHA – 300 forms). However, keeping data in addition to that required for compliance purposes may be desirable in order to:

- track performance and measure progress
- manage risk
- measure the effectiveness of interventions

- objectively identify the areas where the greatest safety concerns exist
- identify costs associated with safety issues in the sense that the data may be used to make risk–based and cost-effective safety decisions in conjunction with the business case for safety.

The scope of this chapter involves an outline of methods for collecting, managing, and analyzing data. Analytical methods will be addressed, although it must be understood that this chapter is not, nor is it intended to be, an exhaustive treatise on analytical methods.

COLLECTION

The first stage of the process is assembly of the data, which may or may not require field data collection. The following must be taken into consideration.

- The purpose or use of the data must be defined before data collection starts.
- It must be determined whether data may be collected from already existing records, or whether the data must be specified and collected without the benefit of any historical records.
- The necessary data fields must be specified (a part of the data management process).
- A standard list of values must be created to avoid different terminologies for the same referent. For example, ensure that the "upper limb" is used consistently and does not appear as "arm" or "hand" or "fractured" does not appear as "broken" or "busted" because in each case the synonyms refer to the same event. This task is also an important aspect of data management. Because terms that are synonymous must be standardized to one term only per event, the use of a mechanism such as a drop-down menu becomes very important to allow data retrieval.
- The mechanism for data collection must be determined (for example, a standardized spreadsheet or a web-based format).

These guidelines are important because inaccurate, corrupted, or incomplete data will seriously reduce the usefulness of the data. To enhance data integrity, it is important to cross-check or to validate the data with different systems, such as human relations, payroll/

financial, medical, or any other entity within the organization that may also collect the data of interest. Finally, any possible limitations of the data must be considered when interpreting the results of any analyses conducted.

Because the consistency of data is important, consider each of the following.

- Success at data collection is dependent upon proper training of individuals involved in each aspect of the data collection process.
- Each entity or business unit must be aware of what data is needed and how the data must be collected and entered into the database.
- The collection and entry process must be convenient, simple, and well communicated.
- There must be a formal data validation process to ensure that the process is working (i.e., that you're getting what you want).
- Reporting from employees (much of the data will deal with accidents, injuries, and illnesses, where "blame" may be perceived) should be elicited through nonpunitive means to encourage early and accurate reporting.

MANAGEMENT

An efficient and useful method to manage the data is to develop an electronic Data Tracking System (DTS). This is an inclusive term that will be used to refer to any system designed to store and manage data.

There are three concerns in establishing a DTS. The first is the determination of what data is to be stored (i.e., the content of the database); the second is the hardware and software to be used to store the data (the infrastructure); and the third is the manner in which the data is to be retrieved for future processing and analysis (data retrieval).

Content of the DTS

The categories of data that are kept in the database are determined by the utility of that data to the organization. However, it must be understood that data beyond that which is normally collected may be required to answer many common questions that arise regarding

safety. Finally, it should be noted that local privacy laws have a considerable impact on data collection and use.

Data for the DTS may be placed in four categories according to data source. These categories are outlined below with examples meant to be illustrative and not exhaustive.

1. **Employee or equipment data.** This category refers to data on the individual employees (such as demographics) or to specific pieces of equipment. The location and availability of this data is highly dependent upon practices in the organization and regulations in the country in which the data is collected.

Example employee data fields:

- unique identifier
- age
- date of hire/time on job or task
- gender
- ethnicity/race
- employee identifier
- employee's department/work location
- employee's occupation or occupation code
- training
- shift
- knowledge of appropriate safe practices (if not only in the case of an event)

Example equipment data fields:

- truck type
- aircraft type
- fuel facilities

Possible source(s):

- human resources

2. **Event data.** This category refers to data that documents specific information regarding the event. This is a more structured approach to organizing the data that may often be present in a narrative. Specific data fields are preferable to a narrative

because data fields ensure the presence of certain data deemed important and allow those fields to be more easily and reliably extracted and subjected to a quantitative analysis.

The quality of initial investigation is a very important aspect of the event data. The data is usable only if it is reliable.

Example data fields:

- body part affected
- International Classification of Disease (ICD) code
- location of incident
- incident identifier (such as an incident number)
- type of incident (e.g., recordable, presence of an ergonomics "flag," etc.)
- plant location of incident
- employee treatment for incident
- number of days restricted
- number of days lost
- date and time
- environmental conditions (weather related, work related)
- time on shift/amount of exposure
- equipment
- materials
- PPE
- tools
- knowledge of appropriate safe practices (if only collected in the case of an event)
- free field for event description
- activity at time of event
- results of medical evaluations

Possible source(s):

- site medical
- site safety
- incident investigator

3. Denominator data. Data in this category is that which allows the calculation of rates. Rates allow comparisons between groups such as different departments, plants, or companies whose exposures are not the same.

Example data field:

- number of hours logged per person per week at each plant in the company
- number of flights/turns
- departures
- number of employees

Possible data source(s):

- payroll
- flight operations

4. Hazard analysis/exposure data. In analysis of safety data, additional risk management may be possible by reviewing or analyzing data on recognized hazards. This allows an accurate assessment of the risk level.

Example data fields:

- job hazard analyses
- hazard tracking system — tracking reports of hazardous conditions (facilities, GSE equipment)
- risk hazard assessment level or index

Possible data source(s):

- site safety
- site ergonomist

The above list of data categories may be used as guides to specify the data fields. Other data fields may be found to be important and relevant, depending on the needs of the organization (e.g., off-duty injuries). This is data that is not normally included in an occupational database but may be seen as pertinent. All data should be linked by at least one field such as an employee identifier. For instance, an accident report should be able to be linked to the employee information, the job hazard analysis, and the exposure data by a common field.

Finally, the extent of the contents of the database must be balanced between need and utility, as well as between the costs and efforts associated with the data collection. This balance is an important policy decision.

DTS Infrastructure

Infrastructure refers to the hardware and software. Hardware consists of shared drives or servers that have sufficient storage space for current and future data. The infrastructural software is that which is used for extracting data, such as query languages. Software may also refer to tools such as Internet-based applications and off-the-shelf data management tools. These are numerous and their use will depend on the budget and needs of the organization.

The hardware and software chosen for data management will usually depend on what is already available. The size of the company and the nature of its existing infrastructure will determine the type of hardware and software that may be added. Note that although current infrastructure may be adequate, it might not allow for expansion and may be more expensive to fix later than to upgrade now. A desktop application may have the advantage of portability, with the disadvantage of size limitation that may not accommodate future growth. However, in some cases a fixed pre-existing infrastructure may be an inherent limitation and should be recognized as preferable to no database.

Choice of infrastructure should also include consideration of the geographic distribution of the entities that must provide data. Therefore, an organization with locations spread over an entire country, or worldwide, should seriously consider a web-based data collection tool. In contrast, a smaller organization that

is located within a small geographical area may find that a local database may be adequate.

Finally, it should be understood that the infrastructure is not a solution. These are simply necessary preparations for data collection that will facilitate collection if the infrastructural needs are considered and consciously decided upon.

Data Input and Extraction (Retrieval)

The method of data input and extraction depends on the infrastructure chosen for data storage. However, the most generic example is represented by the ubiquitous spreadsheet. The most fundamental structure of a data file is to include the data fields (also sometimes called variables) in the columns. As discussed before, typical data fields are injury type, employee identifier, cost, etc. Each case or event is placed in a row (sometimes called a record). Table 20-A shows a typical example of a database with eight variables or data fields, populated by five hypothetical events.

The value of choosing this type of data structure is that most analytical tools are designed to read data in this format. Extracting data from a file with this structure may be as simple as a sort (if one is using Excel, for example) or as complex as a query with a language suited for that purpose. There are many software applications available. Examples used for data analysis are SAS (which includes a statistical program, among many other applications), SPSS (Statistical Package for the Social Sciences), MINITAB, and STATA, among others.

TABLE 20-A. EXAMPLE OF A SIMPLE DATABASE

Employee	Department	Date	Time	Body Part	Accident Type	Sex	Age
1	X	1/2/1997	0700	elbow	struck against	M	50
2	Y	1/3/1997	1500	eye	exposure	M	29
3	Z	1/3/1997	1900	knee	contact with	M	49
4	X	1/4/1997	0630	ribs	struck against	M	53
5	Y	1/7/1997	0930	thumb	struck by	F	41

ANALYSIS

The ultimate purpose of collection and storage of data is to conduct analyses that benefit the organization. The analyses that are possible are only limited by the needs and the available expertise *if the appropriate data has been collected*. Analyses may range from simple descriptive methods that include graphical methods (control charts, pareto charts, etc.), rate calculations, and statistics (mean, median, mode, range, etc.) to more complex inferential methods such as hypothesis tests and odds ratios, among others.

Descriptive Methods
Rate Calculations and Graphical Methods.
One of the most common descriptive techniques is the calculation of rates. For example, the OSHA rate standardizes all data to 100 full-time employees, each working 2000 hours per year, i.e., to 200,000 person-hours. The general formula for rates is:

$$rate = \left(\frac{a}{b}\right)k$$

In this formula, the numerator (a) is a count of the events of interest. For the OSHA rate, this may be recordable accidents. The denominator (b) must be drawn from the total exposed population, such as hours worked or person-hours worked. For the calculation of OSHA rates, this value is person-hours worked at the location or within the entity where the recordable accidents in the numerator were observed. Finally, the quantity (k), which may be referred to as the multiplier, is the standardization value. For the OSHA rate example, this multiplier is 200,000 person-hours. It is important to note that the units for the multiplier (k) and the denominator (b) must be the same.

An example of a rate calculation using the OSHA denominator is to consider 15 injuries among a population of employees who have worked 394,456 person-hours over the course of a calendar year. The OSHA rate is calculated as follows:

$$rate = \left(\frac{a}{b}\right)k = \left(\frac{15}{394,456}\right)200,000 = 7.6 \text{ injuries/100 fte}$$

Figure 20-1 illustrates how rates may be used for the comparison of the recordable injury/illness experiences between several different sectors of U.S. industry as recorded by the U.S. Bureau of Labor Statistics.

A commonly used, simple, and useful technique for understanding data is the pareto chart. A pareto chart requires the data to be ranked from high to low and

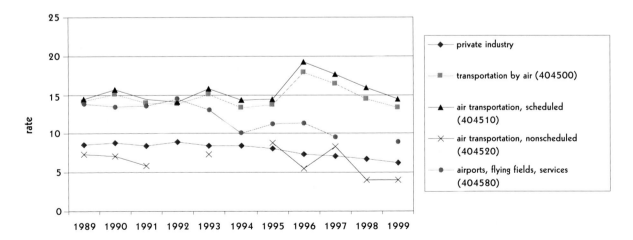

Figure 20-1. Recordable injury/illness cases per 100 full-time employees

U.S. Bureau of Labor Statistics

then graphed, usually as a histogram. There are a number of uses of data placed in a pareto chart.

For instance, in order to better understand slips, trips, and falls, a safety professional reviews the rates of injury types that have been recorded during the previous calendar year at the facility (Figure 20-2).

Next, all cases of slips, trips, and falls are broken down by location of occurrence. In this case, it will be adequate to use frequencies (the term "frequency" is used synonymously with the concept "number of incidents") rather than rates (Figure 20-3).

Finally, the parking lot slips, trips, and falls are broken down by time of day. This is shown in Figure 20-4.

By using the pareto analysis in this manner, the locations and times of slips, trips, and falls may be reviewed to make a qualitative inference about causes of this injury type.

Another example may involve an analysis of manual material handling tasks by frequency (note that this is not preferred but that rates may not be available).

Assume that 43 injuries have been recorded on the ramp. These have been categorized by "type of accident" as shown in Figure 20-5.

It is apparent that the greatest number of accidents occurs in the "material handling" category. However, in this instance, there is already a program in place to address this issue. Therefore, the next most frequent type of accident is "falls," which had not previously been identified as a problem. Therefore, the 12 falls are further categorized by "object involved" in Figure 20-6.

Falls from the belts of the belly loaders account for eight of twelve injuries. Furthermore, a review of these cases showed that seven of the eight occurred during icy or rainy conditions and that the handrails on the belly loaders were not being used.

After this pareto analysis, the final step is to develop an action plan. For this data an action plan may be developed that includes:

1. An audit that concentrates on compliance regarding the use of the handrails on the belly loaders under both wet and dry conditions. This will also

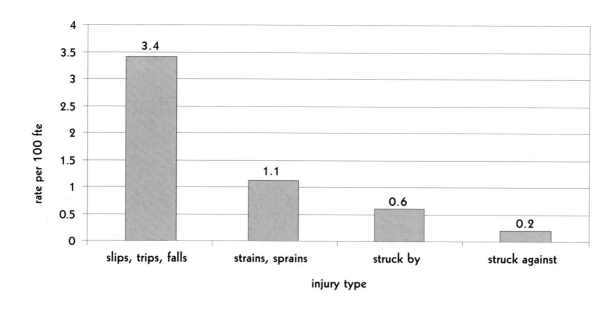

Figure 20-2. Rates (per 100 full-time employees) by injury type arranged in decreasing order.

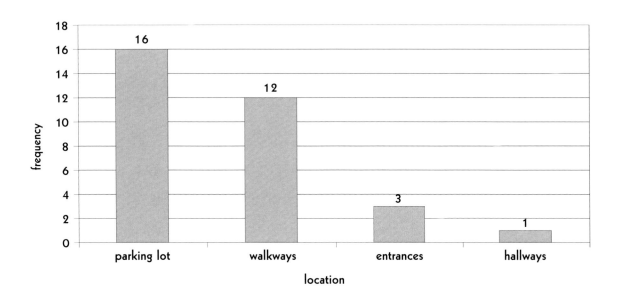

Figure 20-3. Frequency of all cases of slips, trips, and falls by location of occurrence.

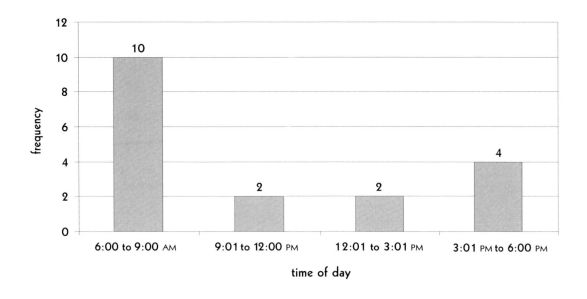

Figure 20-4. Frequency of all cases of slips, trips, and falls by time of occurrence.

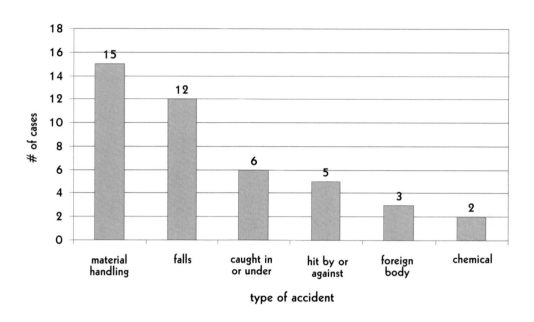

Figure 20-5. Frequency analysis of type of accident.

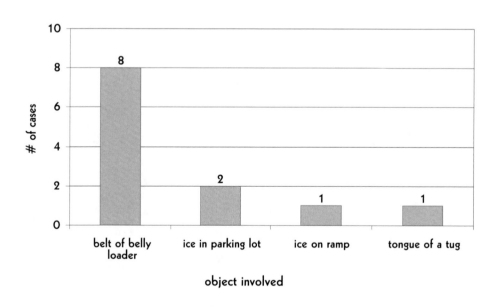

Figure 20-6. Frequency analysis of object involved in the 12 cases of falls.

evaluate the reality of this problem.

2. Training the employees on the importance of using the handrails, a review of the injury history that has resulted from not using the handrails, and that this is an OSHA compliance issue.

3. Stressing enforcement with management and holding them accountable.

4. Reviewing performance to ensure that the action plan has been adhered to and has been successful in changing behavior.

Statistics. Descriptive statistics (as other descriptive methods) help to better understand the data but do not allow the testing of hypotheses. For instance, one may better understand data by knowing the average (mean) and the dispersion (standard deviation and variance). However, these values do not provide answers to questions such as the difference between sub-groups of the data, for example.

Consider the following group of numbers. These numbers are data, but do not provide much information:

98 92 102 89 90 72 69 84 58 65

This data is difficult to interpret because it has been collected, but has not been managed, described, or analyzed. A little data management and labeling may result in the following organization of the data:

heart rate for five males after loading an aircraft:
98 92 102 89 90

heart rate for the same five males 30 minutes later (after recovery): 72 69 84 58 65

Next, descriptive statistics will result in the following summary of the data to help better understand the central tendency (mean, or arithmetic "average") and the dispersion or variability (standard deviation) in each group. This is best described using the mean and standard deviation. These quantities are calculated as follows:

heart rate while loading: mean = 94.2,
standard deviation = 5.6

heart rate after recovery: mean = 69.6,
standard deviation = 9.6

Now that the data has been described, some limited interpretation is possible. However, it is still not possible to answer a specific question (a hypothesis) without the use of an inferential method. An inferential method (or test) would allow the investigator to answer a question such as whether the mean heart rate is statistically significantly lower after recovery than directly after work. (Note that this is, of course, a prosaic question, but one that serves to provide an intuitive basis for understanding the purpose of inferential testing.)

Inferential Methods

As mentioned, inferential methods allow the testing of hypotheses and the evaluation of statistical significance. This covers a wide variety of techniques that are outside of the scope of this chapter. However, a few of the more commonly used techniques are:

- hypothesis tests
- regression
- analysis of variance (ANOVA)
- odds ratios and relative risks (epidemiological methods)

The odds ratio will provide a good example of a typical inferential method that may be used with observational data (as are often encountered in industry). Assume that the data in Table 20-2 represents cross-sectionally observed data. This data is arranged in a 2-by-2 contingency table representing the cases of carpal tunnel syndrome among 1,104 employees exposed to an assembly task. For comparison, 846 employees were located who work under similar conditions but who do not perform the assembly task. It can be seen that 21 of these employees also developed carpal tunnel syndrome even though they had no exposure to the task.

The most appropriate method for analyzing the data shown in Table 20-2 is to calculate the odds ratio (OR) to determine the effects of exposure to the assembly tasks:

$$OR = \frac{ad}{bc} = \frac{56 * 825}{1048 * 21} = 2.1$$

This result means that those employees who were exposed to the hypothetical assembly task were found to be more than two times more likely to develop carpal tunnel syndrome than those who were not exposed.

A discussion of more advanced inferential statistical methods and the idea of statistical significance are beyond the scope of this chapter. However, an outstanding introduction to these methods may be found in Hines *et al.* (2003).

CONCLUSION

The first step in the process of data collection and analysis is to determine what the ultimate use of the data will be. This has to be determined and understood before the process of collection, management, and analysis is begun because it will be the limiting factor in all future use and usefulness of the data. Further analyses may also be conducted beyond those initially envisioned, given that the data is suited to these analyses.

Collecting safety data is useful in providing guidance for decisions that will result in event reduction and cost reduction associated with workman's compensation, equipment repair, etc. A data tracking system is an important and necessary function for successful organizations. However, statistical analysis is difficult to do if the practitioner is not motivated or trained in its collection and presentation.

REFERENCES

Statistics and Epidemiology

Checkoway, H., Pearce, N., and D. Crawford-Brown. *Research Methods in Occupational Epidemiology.* New York: Oxford University Press, 1989.

Gordis, L. *Epidemiology, 3rd Edition.* Philadelphia: Elsevier Saunders, 2004.

Hines, H.W., Montgomery, D.C., Goldsman, D. M., and C.M. Borror. *Probability and Statistics in Engineering, 4th Edition.* New York: Wiley, 2003.

Hogg, R.V., and E.A. Tanis. *Probability and Statistical Inference, 6th Edition.* Upper Saddle River: Prentice Hall, 2001.

Kelsey, J., Whittemore, A., Evans, A., and D. Thompson. *Methods in Observational Epidemiology, 2nd Edition.* New York: Oxford University Press, 1996.

Pagano, M., and K. Gavreau. *Principles of Biostatistics, 2nd Edition.* Pacific Grove, CA: Duxbury Thomson Learning, 2000.

Wheeler, D. J. *Understanding Variation: The Key to Managing Chaos, 2nd Edition.* New York: SPC Press, 1999.

TABLE 20-2.
EXAMPLE OF DATA STRUCTURED FOR THE CALCULATION OF AN ODDS RATIO

		carpal tunnel syndrome		
		yes (case)	no (control)	
exposed to assembly task	yes	a=56	b=1048	1104
	no	c=21	d=825	846
		77	1873	1950

Safety Analysis

Brauer, R.L. *Safety and Health for Engineers.* New York: Wiley, 1994.

Harms-Ringdahl, L. *Safety Analysis: Principles and Practice in Occupational Safety, 2nd Edition.* New York: Taylor and Francis, 2001.

Roland, H.E., and B. Moriarty. *System Safety Engineering and Management, 2nd Edition.* New York" Wiley, 1990.

Walters, H.A. *Statistical Tools of Safety Management.* New York: Van Nostrand Reinhold, 1995.

Web sites for Industry Data

United Sates Bureau of Labor Statistics Public Data Query: http://data.bls.gov/PDQ/outside.jsp?survey=sh.

National Safety Council Research and Statistics: http://www.nsc.org/lrs/statstop.htm.

WORKING GROUP MEMBERS

Co-Chairmen:

Maxwell Fogleman, Ph.D., MPH – Embry-Riddle University

John Kane – Air Canada

Members:

Ken Roberts – Frontier Airlines, Inc.

Carlos Romero – LSG Skychefs

Rob Owen – BA Aviation Shared Services, Inc.

John Phillips – United States Air Force

Jason Schlattman – America West Airlines

Michael Lueck – ABXAir Inc

Contributors:

Colin Hewett – BBA Aviation Shared Services Inc.

Cathy Hollister – Air Canada

Rich Knight – Air New Zealand

Choon Woo Lee – Korean Air

Melvyin Lim – Singapore Airlines

Kim McDaniel – Southwest Airlines

John Montgomery – Airserv

Mark Paterson – Royal Australian Air Force

Tim Racicot – Continental Airlines

Fred Rose – Air Wisconsin

Robert Schwartz – Air Canada

Bill Simpson – Piedmont Hawthorne